Are You Prepared?

Legal Issues Facing North Carolina Public Employers in Disasters and Other Emergencies

2014

Diane M. Juffras

UNC
SCHOOL OF
GOVERNMENT

The School of Government at the University of North Carolina at Chapel Hill works to improve the lives of North Carolinians by engaging in practical scholarship that helps public officials and citizens understand and improve state and local government. Established in 1931 as the Institute of Government, the School provides educational, advisory, and research services for state and local governments. The School of Government is also home to a nationally ranked graduate program in public administration and specialized centers focused on information technology and environmental finance.

As the largest university-based local government training, advisory, and research organization in the United States, the School of Government offers up to 200 courses, webinars, and specialized conferences for more than 12,000 public officials each year. In addition, faculty members annually publish approximately 50 books, manuals, reports, articles, bulletins, and other print and online content related to state and local government. Each day that the General Assembly is in session, the School produces the *Daily Bulletin Online*, which reports on the day's activities for members of the legislature and others who need to follow the course of legislation.

The Master of Public Administration Program is offered in two formats. The full-time, two-year residential program serves up to 60 students annually. In 2013 the School launched MPA@UNC, an online format designed for working professionals and others seeking flexibility while advancing their careers in public service. The School's MPA program consistently ranks among the best public administration graduate programs in the country, particularly in city management. With courses ranging from public policy analysis to ethics and management, the program educates leaders for local, state, and federal governments and nonprofit organizations.

Operating support for the School of Government's programs and activities comes from many sources, including state appropriations, local government membership dues, private contributions, publication sales, course fees, and service contracts. Visit www.sog.unc.edu or call 919.966.5381 for more information on the School's courses, publications, programs, and services.

Michael R. Smith, DEAN
Thomas H. Thornburg, SENIOR ASSOCIATE DEAN
Frayda S. Bluestein, ASSOCIATE DEAN FOR FACULTY DEVELOPMENT
L. Ellen Bradley, ASSOCIATE DEAN FOR PROGRAMS AND MARKETING
Johnny Burleson, ASSOCIATE DEAN FOR DEVELOPMENT
Todd A. Nicolet, ASSOCIATE DEAN FOR OPERATIONS
Bradley G. Volk, ASSOCIATE DEAN FOR ADMINISTRATION

FACULTY

Whitney Afonso
Trey Allen
Gregory S. Allison
David N. Ammons
Ann M. Anderson
Maureen Berner
Mark F. Botts
Michael Crowell
Leisha DeHart-Davis
Shea Riggsbee Denning
Sara DePasquale
James C. Drennan
Richard D. Ducker
Joseph S. Ferrell
Alyson A. Grine
Norma Houston

Cheryl Daniels Howell
Jeffrey A. Hughes
Willow S. Jacobson
Robert P. Joyce
Diane M. Juffras
Dona G. Lewandowski
Adam Lovelady
James M. Markham
Christopher B. McLaughlin
Kara A. Millonzi
Jill D. Moore
Jonathan Q. Morgan
Ricardo S. Morse
C. Tyler Mulligan
Kimberly L. Nelson
David W. Owens

LaToya B. Powell
William C. Rivenbark
Dale J. Roenigk
John Rubin
Jessica Smith
Meredith Smith
Carl W. Stenberg III
John B. Stephens
Charles Szypszak
Shannon H. Tufts
Vaughn Mamlin Upshaw
Aimee N. Wall
Jeffrey B. Welty
Richard B. Whisnant

© 2014
School of Government
The University of North Carolina at Chapel Hill

Printed in the United States of America

18 17 16 15 14   1 2 3 4 5

ISBN 978-1-56011-753-7

♾ This publication is printed on permanent, acid-free paper in compliance with the North Carolina General Statutes.
♲ Printed on recycled paper

Contents

PART 3
Issues Specific to Public Health Emergencies   49

Conclusion   81

Introduction

North Carolina public employers face natural and man-made disasters, ranging from hurricanes, tornadoes, severe winter storms, and flooding to pandemic flu and the possibility of bioterrorism. Some of these emergencies have already occurred; others have not. And new threats arise all the time—building fires accompanied by the release of hazardous gases, chemical spills on roadways or into the environment, and the possibility of major disruptions to the power grid. Never is the demand for government services as great as in emergency situations such as these. Yet providing essential services is a daunting challenge, not only because of physical and environmental obstacles, but because government employees and their families must deal with the same issues facing the population at large. Thanks in no small part to the dedication of North Carolina's public employees, state and local governments have maintained the continuity of essential services despite these types of hardships and uncertainty.

This book discusses the employment law issues faced by governmental employers that are specific to disasters and other emergencies. Part 1 focuses on preparing for disaster and emergency circumstances generally. Part 2 addresses compensation issues likely to arise when some employees must work many additional hours, while others cannot or will not make it to the workplace. Part 3 considers the legal issues that will be specific to public health emergencies such as an outbreak of pandemic flu or the release of a biological agent by terrorists.

Under the North Carolina Emergency Management Act, an *emergency* is "an occurrence or imminent threat of widespread or severe damage, injury, or loss of life or property resulting from any natural or man-made accidental, military, paramilitary, weather-related, or riot-related cause."[1] The governor of North Carolina may declare a state of emergency either statewide or in particular geographic areas. A *disaster declaration* is "a gubernatorial declaration that the impact or anticipated impact of an emergency constitutes a disaster of one of the types enumerated in G.S. 166A-19.21(b)." The types of disaster recognized under Section 166A-19.21(b) of the North Carolina General Statutes (hereinafter G.S.) are distinguished from one another by whether a federal disaster declaration has been issued and whether the disaster is severe enough for the federal government to provide financial assistance to the area. In addition to the governor's authority to declare a state of emergency, cities and counties may also declare states of emergency pursuant to G.S. Chapter 166A.[2]

Except where specifically noted, however, the discussions in this book apply to any situation in which a North Carolina governmental employer faces the actual or imminent threat of widespread or severe damage, injury, or loss of life or property, regardless of whether an official declaration has been made that such an emergency exists. When the situation is bad, the official designation is less important than the public agency's response.

---

1. *See* N.C. GEN. STAT. (hereinafter G.S.) § 166A-19.3(6).
2. *See* G.S. 166A-19.22. The General Assembly also has the authority to declare a state of emergency under G.S. 166A-19.20.

Part 1

Preparing for a Disaster

Certain employment issues will arise regardless of the particular emergency, and human resource managers should prepare for them in advance. Such preparations should include establishing an employee evacuation plan that takes into account disabilities and other limitations individual employees may have; developing backup chains of command in case of the absence or incapacity of key members of the management team; and planning a response to increased rates of absenteeism that includes reassigning employees, calling upon mutual aid agreements, supplementing the workforce through the use of temporary employees or volunteers, or facilitating telecommuting.

Having an Evacuation Plan That Takes into Account Employees with Disabilities

As a general rule, the Americans with Disabilities Act (ADA) prohibits employers from asking employees questions about a medical condition unless there is reason to believe that a disability is interfering with an employee's performance of his or her job duties. There is one situation, however, in which it is in both the employee's and the employer's interest to have an accurate identification of employees with special needs: where a natural disaster, terrorist attack, or other public health emergency requires an evacuation of the workplace.

An employer may not exclude an employee with a disability from the workplace at any time—even in anticipation of a public emergency—unless that employee poses a direct threat to others. Thus, an employer who fears or has reason to believe that a situation requiring an evacuation may develop may not require employees with disabilities to stay home to make the evacuation easier. An employer can and should, however, survey its employees to determine who will require assistance of what kind in an emergency. The key to surveying employees about their medical needs in a lawful manner is to explain the reason for the inquiry and to make clear that revealing information about the existence of a disability is entirely voluntary. An employer may, of course, ask employees with known disabilities if they will require assistance. As part of this survey, an employer may also ask employees who indicate a need for assistance because of a medical condition to describe the type of assistance they think they will need:

> One way that this can be done is by giving all employees a memo with an attached form requesting information. The employer also may have a follow-up conversation with an individual when necessary to obtain more detailed information. For example, it would be important for an employer to know whether someone who uses a wheelchair because of mobility limitations is able to walk independently, with or without the use of crutches or a cane, in an emergency situation. It also would be important for an employer to know if an individual will need any special medication, equipment, or device (*e.g.*, an assisted wheelchair carrier strap or a mask because of a respiratory condition) in the event of an emergency. Of course, an employer is entitled only to the information necessary for it to be prepared to provide assistance. This means that, in most instances, it will be unnecessary for an employer to know the details of an individual's medical condition.[1]

Information obtained through a survey of this kind must remain confidential and may be disclosed only to those with responsibilities for emergencies. It should not be disclosed to supervisors unless they will have direct responsibility for the evacuation of such employees.

---

1. *See* U.S. Equal Employment Opportunity Commission (EEOC), *Fact Sheet on Obtaining and Using Employee Medical Information as Part of Emergency Evacuation Procedures,* www.eeoc.gov/facts/evacuation.html.

Pre-Screening Employees Who Will Perform Natural Disaster Cleanup Work

The Occupational Safety and Health Act (OSHA) requires employers to provide employees a workplace free from recognized hazards likely to cause death or physical harm.[2] The act also requires employers to provide employees personal protective equipment to protect against known workplace dangers. Finally, the act's Bloodborne Pathogens Standard requires employers to have a written exposure control plan detailing how they will prevent or minimize first responders' exposure to blood and other bodily fluids potentially infected with hepatitis B virus, hepatitis C virus, and the human immunodeficiency (HIV) virus.[3] Neither the federal nor the state government has authority to suspend these or other occupational health and safety standards in the case of natural disaster. But working during or in the aftermath of a natural disaster or a public health emergency presents hazards to employee health that may not be found in the course of an employee's regular work. The most frequent hazards associated with disaster response activities are sharp, jagged debris; floodwater exposure; electrical hazards; slick and unstable surfaces; and contact with blood or bodily fluids and with human and animal remains.[4] The possibility that employees may be exposed to hazardous conditions or hazardous chemicals or agents may or may not be predictable. Whether these conditions or agents can be controlled may or may not be predictable.

For these reasons the National Institute of Occupational Safety and Health (NIOSH), a division of the U.S. Centers for Disease Control and Prevention (CDC), recommends that government employers perform pre-deployment or pre-exposure medical screening of workers who will assist in natural disaster work. This is because it is difficult to assess the health effects that disaster work has had on an employee without a baseline from which to evaluate the person's health. The agency also urges employers to

---

2. *See* 29 U.S.C. § 654(a). Occupational Safety and Health Act standards are set out at 29 C.F.R. Part 1910.

3. *See* National Institute of Occupational Safety and Health (NIOSH), Centers for Disease Control and Prevention (CDC), Publication Number 2008-115, *Protect Your Employees with an Exposure Control Plan*, www.cdc.gov/niosh/docs/2008-115.

4. *See* NIOSH, CDC, *Hazard Based Guidelines: Protective Equipment for Workers in Hurricane Flood Response* (hereinafter *Protective Equipment for Workers in Hurricane Flood Response*), www.cdc.gov/niosh/topics/emres/pre-workers.html.

conduct post-exposure screening of employees who have been working in disaster recovery areas. NIOSH also recommends pre-screening to assess an employee's fitness safely to perform hazardous or stressful work that is different than the work he or she usually performs.[5]

NIOSH recommends that employers also assess in advance employees' ability to use personal protective equipment necessary for the tasks they will perform.[6] This may involve fitting employees likely to be reassigned to disaster work with personal protective equipment the employer already has or can easily obtain. Some forms of emergency, however, will require less familiar protective gear. In the event of a terrorist attack involving chemical, biological, radiological, or nuclear agents, employers will also be responsible for supplying first responders with appropriate personal protective equipment as identified by the Occupational Safety and Health Administration, NIOSH, and the Department of Homeland Security.[7] An effective response to a disaster requires that employers address issues of worker health and safety in advance.

---

5. *See* NIOSH, CDC, *Guidance for Pre-exposure Medical Screening of Workers Deployed for Hurricane Disaster Work*, www.cdc.gov/niosh/topics/emres/preexposure.html. *See also* NIOSH, CDC, *Guidance for Post-exposure Medical Screening of Workers Leaving Hurricane Disaster Recovery Areas*, www.cdc.gov/niosh/topics/emres/MedScreenWork.html. The National Response Team has prepared a technical assistance document, *Emergency Responder Health Monitoring and Surveillance*, that details best practices in medical monitoring of first responders during the course of emergency responses of various types and sizes. It is available at http://nrt.sraprod.com/ERHMS/ and through links at http://blogs.cdc.gov/niosh-science-blog/2012/09/responderhealth/.

6. For guidelines on providing appropriate personal protective equipment for employees responding to disasters, see *Protective Equipment for Workers in Hurricane Flood Response.*

7. *See* 29 C.F.R. §§ 1910.120(g) and 1910.156(e). See also U.S. Department of Labor, OSHA, *OSHA/NIOSH Interim Guidance: Chemical-Biological-Radiological-Nuclear (CBRN) Personal Protective Equipment Selection Matrix for Emergency Responders* (April 2005), www.osha.gov/SLTC/emergencypreparedness/cbrnmatrix/index.html.

Anticipating Employee Absences Following a Disaster or Emergency

Public employers should assume two important points during a public emergency, whether the emergency is a natural disaster or a public health emergency. First, employers should assume there will be an increased demand for government services—public safety, public works, and public health especially. Second, employers should assume there will be increased absenteeism because employees or their immediate family members are injured or ill, have lost their homes, have no transportation, have lost access to child or elder care, or refuse to come to work. Employers should therefore take a pre-emergency survey to see where they are likely to be short-staffed and to plan in advance how they will alleviate that shortage.

Surveying Employees

In surveying employees, an employer may not ask an employee to disclose if he or she has a compromised immune system or chronic health condition that could make that employee more susceptible to complications of a communicable disease in circulation, such as pandemic flu, or an agent released by terrorists, such as smallpox or anthrax. The answer would likely disclose the existence of a disability. Such a question is impermissible in the absence of objective evidence that such complications and their symptoms will cause a direct threat to others.

Employers may, however, make inquiries that are not related to a disability. An inquiry is not disability-related if it is as likely to elicit non-medical reasons for absence during a disaster (for example, limited public transportation) as medical reasons (such as chronic illnesses that increase the risk of complications). The survey should be structured so that employees may give a yes or no answer to the entire question without specifying the factors that apply to them.[8] A survey prepared by the U.S. Equal Employment Opportunity Commission that may be given to employees to predict the rate of absenteeism in advance of an actual emergency is reproduced below. The survey does not ask employees to state the reason they are likely to be absent from work, only that they would be unable to get to work.

---

8. *See* EEOC, *Pandemic Preparedness in the Workplace and the Americans with Disabilities Act* (hereinafter *Pandemic Preparedness in the Workplace*), www.eeoc.gov/facts/pandemic_flu.html.

ADA-Compliant Pre-pandemic Employee Survey

<u>Directions:</u> Answer "yes" to the whole question *without specifying the factor that applies to you.* Simply check "yes" or "no" at the **bottom of the page.**

In the event of a pandemic, would you be unable to come to work because of any one of the following reasons:

- If schools or day-care centers were closed, you would need to care for a child;
- If other services were unavailable, you would need to care for other dependents;
- If public transport were sporadic or unavailable, you would be unable to travel to work;
- and/or;
- If you or a member of your household fall into one of the categories identified by the CDC as being at high risk for serious complications from the pandemic influenza virus, you would be advised by public health authorities not to come to work (e.g., pregnant women; persons with compromised immune systems due to cancer, HIV, history of organ transplant or other medical conditions; persons less than 65 years of age with underlying chronic conditions; or persons over 65).

Answer: YES _____, NO _____ [9]

Preparing an Emergency Chain of Command and Lists of Essential Employees

To respond quickly to the unique demands of a natural disaster, public health emergency, or terrorist attack, a unit of local government should have at the ready an emergency chain of command specific to the particular circumstances and a list of essential personnel needed to respond. Local government employers in compliance with the Federal Emergency Management Agency (FEMA) National Incident Management System (NIMS) will have a chain of command for each kind of potential emergency situation and a list of essential personnel in place. Those who do not should start preparing these fundamental response documents now.

---

9. *See Pandemic Preparedness in the Workplace.*

Since 2004 all public agencies have had to comply with NIMS in order to receive federal preparedness funding.[10] How local governments manage their normal, day-to-day operations rarely suffices under emergency conditions, where timeliness is critical. NIMS is designed to provide a common standard for overall incident management for all response agencies in the country, no matter where, no matter what the unit of government, no matter what kind of incident. Under NIMS, lines of control related to emergency operations are organized along task-oriented lines rather than according to the normal departmental structure governments use to accomplish long-term goals or to support everyday services. At the head of the NIMS command staff is the Incident Commander, or IC, who is the overall incident manager. This may be a single person, who may or may not have a deputy. In a city or county, the IC may not be the city or county manager. It may be the emergency management director, the police chief or sheriff, or the fire chief. In a Unified Command, the IC position may be held by two or more ICs from different response entities, who set the objectives for the incident and co-manage it.[11]

NIMS does not take command away from state and local authorities, nor does it supplant the normal lines of final authority within a city, county, or other local agency. For example, even if the fire chief is the IC for a particular emergency, the city or county manager still retains hiring and firing authority for employees. Similarly, if the county emergency management director is the IC for a bioterror incident involving the release of a chemical agent and county health department personnel are under the emergency management director's command, the county health director would still maintain ultimate personnel authority over those employees. NIMS simply provides the framework to enhance the ability of responders to work together more effectively. As FEMA explains it, "Entities that have integrated NIMS into their planning and incident management structure can arrive at an incident with little notice and still understand the procedures and protocols governing the response, as well as the expectations for equipment and personnel. NIMS provides commonality in preparedness and response efforts that allow

---

10. *See* Homeland Security Presidential Directive 8 (Dec. 17, 2003); U.S. Department of Homeland Security, *National Incident Management System* (Dec. 18, 2008) (hereinafter *NIMS Document*), www.fema.gov/pdf/emergency/nims/NIMS_core.pdf.

11. See *NIMS Document*, at 45–62.

diverse entities to readily integrate and, if necessary, establish unified command during an incident."[12]

Communicating the Organization and Assignment of Responsibilities

Employees will need to know the emergency command structure in advance, especially as it may be different in different emergencies. NIMS-compliant local governments may have already put such emergency command structures in place, but for them to be effective employees must know beforehand how they will work. The plan should specify what will happen if a key employee, such as the city or county manager, the emergency management director, police chief, sheriff, fire chief, or public health director, becomes seriously ill or injured or dies during the course of the disaster and response, and employees should be informed in advance of the order of interim emergency succession. The governing board should consider whether to give the manager or IC the authority to waive local personnel ordinances or policies, and if so, which ones.

Deciding and Communicating Which Services Are Essential

Local governments must decide which services to focus on in which kinds of disaster and what the threats to maintaining those services are likely to be. In a natural disaster, physical damage to infrastructure is likely to cause an interruption of services, both because key components of that infrastructure—such as roads and transportation—are damaged and because that damage makes it difficult for employees to report to work. During a public health emergency such as pandemic flu, as many as 10 to 25 percent of the local government workforce could be absent. Employers should decide in advance, with respect to each type of emergency, which positions are essential, which nonessential personnel can be reassigned to help deliver essential services, and which personnel and services can be temporarily suspended and which departments temporarily closed. Common sense dictates that law enforcement, fire protection, emergency medical services, emergency management, public health services, telecommunications, transportation, and utilities must be maintained in virtually all circumstances.

---

12. *See* U.S. Department of Homeland Security, *NIMS: Frequently Asked Questions*, www.fema.gov/pdf/emergency/nims/nimsfaqs.pdf.

Maintaining Essential Services in the Face of Worker Absences and Shortage

Depending on the nature and the severity of a disaster, some employees will not be able to work. They may be ill or a family member may be, roads may be impassable or vehicles destroyed, dependent care may not be available, or the workplace itself may be destroyed. In some situations, even if a public employer's entire workforce reports for duty, the increased demand for government services the emergency calls forth will leave the jurisdiction short-handed. A government agency can respond to a shortage of workers in a variety of ways. These include reassignment of current employees, letting employees telecommute or work remotely, hiring temporary employees, using volunteers, using retired employees, engaging independent contractors, and accepting help from other jurisdictions with which a public employer has a mutual aid agreement.

Reassignment of Current Employees

Some understaffed but crucial services may be delivered by reassigning "non-essential" personnel to perform the services of "essential" personnel. It is a general rule of employment law that an employer may order an employee to work outside the regular position description so long as the employee is physically and mentally able to do so safely. Indeed, most job descriptions include a requirement that employees perform "other duties as required." Employers should therefore determine in advance which nonessential employees will be reassigned to perform essential services in each type of emergency situation.

As a matter of law, an employee's pay does not have to change. An employer who temporarily reassigns an employee to different work duties during the course of an emergency does not have to adjust that employee's rate of pay, regardless of whether the employee is exempt or nonexempt. The employer may pay the employee at a higher rate if it chooses. On the other hand, the costs of the emergency response might make paying the higher rate fiscally impossible. Overtime rules will still apply. As discussed in more detail below in the section on Fair Labor Standards Act issues, there is no emergency exception to FLSA overtime rules. A nonexempt employee who physically works more than 40 hours in the workweek must be paid overtime or credited with compensatory time off,[13] even if the nonexempt employee is

---

13. *See* 29 C.F.R. § 778.103.

temporarily assigned exempt job duties. If an exempt employee is temporarily assigned nonexempt duties, however, the employer has no duty to pay overtime for hours worked over 40 or any additional compensation at all.[14] This contrasts with the non-emergency situation in which an exempt employee takes a regularly scheduled, second job that is nonexempt. In that case, the employee is not paid overtime but is paid at the regular straight-time rate for the second position for all hours worked in that capacity.[15]

Telecommuting or Working Remotely

Employers should determine in advance of a disaster or a public health quarantine whether they will require some groups of employees to telecommute. Employers should also decide whether they will allow nonessential employees to telecommute in lieu of taking accrued paid leave, or unpaid leave if paid leave is not available. Essential personnel whose job responsibilities may be performed from a remote site may be ordered to work from home. Refusal to do so is a form of insubordination for which an employer may take disciplinary action, up to and including discharge. An employer may also direct an employee who refuses to telecommute either to take unpaid leave or to use accrued paid leave, as the employer prefers.

Having employees work off-site presents a number of challenges. Some of those challenges are more properly the domain of the information technology department than of the human resources and legal departments, such as whether the organization's and individual employee's remote work capabilities are sufficient to meet the organization's needs in this particular situation and whether the organization's information technology capabilities will be sufficient to meet a likely increase in remote usage. Other issues must be addressed by the government employer's manager, agency director, human resources department, and attorney. So, for example, the employer will need to determine in advance how employees will record remote work time and will need to instruct employees in the method chosen. Closely related to the issue of recording work time is determining which employees will be able to work full time from an off-site location. In a disaster, employees may have ill or injured family members or young children who need tending at

---

14. *See* 29 C.F.R. § 541.700.

15. *See* 29 C.F.R. §§ 541.700; 541.604(a). *See also* U.S. Department of Labor Wage and Hour Opinion Letter No. 2005-14, dated March 17, 2005, www.dol.gov/whd/opinion/FLSA/2005/2005_03_17_14_FLSA_nonexempt.pdf.

home. Those responsibilities may make full-time remote work difficult or impossible.

Employers may order or approve telecommuting with or without the use of formal written agreements. Regardless of the approach a public employer takes, supervisors should make clear in advance to employees designated as telecommuters what the remote worksite schedule will be, how the supervisor will know the employee is "at work" at any given time (through use of electronic or telephonic sign-in and sign-out procedures, for example), how the supervisor will be kept informed of the progress of assignments and daily tasks, what technologies the employee is expected to use to maintain contact, and how the employee can access assistance for technical advice or for substantive help on a project.

Employers should clearly communicate to telecommuting nonexempt employees the organization's policies about working through meal periods, rules about phone call and email responses outside of scheduled work hours (as this time constitutes compensable time under the FLSA), and whether overtime must be approved. Employers should have regular policies regarding these matters, but it will be important to remind employees of these polices or inform employees of any policy changes necessitated by an emergency. The bottom line on paying nonexempt employees is that if an employee says he or she has worked and the employer cannot prove otherwise, the time the employee claims as time worked must be paid.

Temporary Employees

Hiring temporary employees is another way to deal with a shortage of qualified workers to handle a local government's unique needs during an emergency. An employer may hire temporary workers in one of two ways: by contracting with an independent business that hires and then places its employees in temporary assignments with organizations with a short-term need for additional help, or by recruiting and hiring employees on a temporary basis on its own.

Using a Temp Agency

Where a temporary agency can supply workers with the kinds of skills that local government employers need in an emergency, contracting may be an efficient choice because the temp agency, and not the employer, interviews the worker and determines if the applicant has the necessary skills and credentials, pays the worker wages, and provides benefits and workers' compensation insurance. Temp agencies generally charge a premium for employees because some of the amount they charge employers goes to covering the agency's overhead costs and making a profit. Nevertheless, a temporary employment agency is likely to be the fastest way to "staff up" in response to an unforeseen need for more employees in particular fields.

Even if a government employer contracts with an agency for hiring and compensating temporary workers, the government employer will nonetheless retain some liability in connection with the workers' presence at the employer's worksite. The government employer will typically control working conditions, provide day-to-day assignments and supervision, and determine the length of the assignment. Under Title VII and other federal antidiscrimination laws, an employer with such control will bear responsibility as a joint employer for any discrimination or harassment to which a worker is subjected while at the local government worksite.[16] A public employer contracting with a temporary staffing agency for short-term workers should therefore ensure that those workers receive copies of any employer workplace policies, identify persons both at the employer's worksite and at the staffing agency to whom a temporary worker should report discrimination or harassment, and instruct supervisors to treat temporary workers with the same respect as regular employees.

Hiring Temporary Employees Directly

Most public employers have experience hiring temporary employees. The need for the swift hiring of potentially large numbers of temporary employees in the case of an emergency, however, dictates that public employers consider a number of issues. For example, many local government employers have policies, some adopted by ordinance and some by resolution, that require that positions be advertised in particular places for particular periods of

---

16. *See, e.g.,* EEOC, *Enforcement Guidance: Application of EEO Laws to Contingent Workers Placed by Temporary Employment Agencies and Other Staffing Firms,* Notice 915.002 (Dec. 3, 1997), www.eeoc.gov/policy/docs/conting.html; EEOC, *Enforcement Guidance: Application of the ADA to Contingent Workers Placed by Temporary Agencies and Other Staffing Firms,* Notice 915.002 (Dec. 22, 2000), www.eeoc.gov/policy/docs/guidance-contingent.html; EEOC, *Questions and Answers: Enforcement Guidance: Application of the ADA to Contingent Workers Placed by Temporary Agencies and Other Staffing Firms* (Dec. 27, 2000), www.eeoc.gov/policy/docs/qanda-contingent.html.

time before applications will be reviewed. Some have nepotism policies that restrict the hiring of persons who have specified family relationships with current employees. Almost all have minimum qualifications set forth for every job included in the position classification system. Employers need to determine in advance (1) whether, as currently written, these policies apply to temporary employees and (2) whether the jurisdiction wants to adopt a policy defining the circumstances in which it will suspend some or all policies of this nature.

Public employers do not, as a matter of federal or state law, have to extend employment benefits to truly temporary employees. The Local Governmental Employees' Retirement System (LGERS) requires that participating jurisdictions enroll an employee in any regular position that requires at least 1,000 hours of service per year.[17] Although neither the General Statutes nor the regulations governing LGERS (issued by the Office of the State Treasurer) define *regular position*, positions authorized for a limited term would probably not qualify, even if the employee worked in excess of 1,000 hours due to the needs of an emergency response. But an individual who is hired on a temporary basis during an emergency and who is later kept on as a permanent employee in a position that requires 1,000 hours each year should receive retirement system credit for the service performed in the temporary position.

Temporary employees hired in response to emergency conditions whose employment does not extend beyond three months do not have to be offered health insurance coverage under the employer's group health plan. The Patient Protection and Affordable Care Act (the "Affordable Care Act," sometimes referred to colloquially as the health care reform law) will require all employers with 50 or more full-time equivalents[18] to offer health benefits that meet a defined standard to their full-time employees and to pay a set portion of the cost of those benefits. The Affordable Care Act defines full-time employees as those who work 30 or more hours per week for three or more months. If a temporary employee is still performing work for an employer after three months, the employer will have to offer health insurance coverage to that employee regardless of whether the position will end within a period of a few more months.

---

17. *See* 20 N.C. Admin. Code 02C.0802.
18. I.R.C. § 4980H(c)(2).

Independent Contractors

In responding to an emergency, local governments may contract with outside companies to provide services such as debris removal, hazard mitigation, construction, or the repair of technology infrastructure—to name just a handful of possible services they may need. When a public employer contracts with an outside company to provide services, the worker providing the service is still an employee of the outside company: the worker's wages, employment taxes, and benefits are paid by the company. This is true even though in some circumstances the public employer makes the day-to-day assignments and supervises the worker.

Sometimes, however, public employers contract with outside individuals to provide services. Because both the public employer and the individual expect the relationship to be temporary, there is a temptation to say the individual has been engaged as an independent contractor (sometimes referred to colloquially as a contract employee). The independent contractor relationship is attractive to employers because independent contractors typically perform their services for a set fee and are not provided with employee benefits or any other form of compensation, such as overtime. In addition, fees paid to independent contractors are not subject to employment taxes—such as FICA (Federal Insurance Contributions Act) or FUTA (Federal Unemployment Tax Act) taxes—to which the employer must make a contribution.

Employers can quickly get into trouble this way. The independent contractor relationship is a distinct legal status. Public employers should be aware that very few work relationships meet the test for independent contractor status. Employers who misclassify workers as independent contractors may incur significant (and unbudgeted) liabilities, such as back overtime, IRS penalties, and the value of lost benefits. The rules of the Internal Revenue Service and U.S. Department of Labor that govern independent contractor status are not waived under disaster or emergency conditions. Given that government entities incur significant expenses during public emergencies, they can probably least afford to incur liabilities associated with worker misclassification at such a time.

In brief, to determine whether a worker is an employee or an independent contractor for the Internal Revenue Code's tax withholding, Social Security, and Medicare contribution purposes, the courts use a common-law test generally known as the right-to-control test. For FLSA overtime purposes, the courts use a version of the right-to-control test called the economic reality test.[19] Under both the right-to-control and economic reality tests, the essence of the relationship between a hiring organization and an independent contractor is the agreement by the independent contractor to do a discrete job according to the independent contractor's own judgment and methods, without supervision by the hiring organization.

There are six general characteristics of an independent contractor:

1. The hiring organization has relatively little control over how an independent contractor does the job.
2. An independent contractor supplies the materials or equipment needed for the job or directly hires others to assist him or her in performing the work.
3. The independent contractor is not paid an hourly wage but retains the ability to make a profit or sustain a loss on the job.
4. An independent contractor usually has a special skill and exercises initiative in seeking out assignments or clients.
5. An independent contractor relationship is usually for a limited duration.
6. An independent contractor usually performs work that is peripheral, rather than essential, to the hiring organization's operations.

If a public employer engages an individual to perform services in connection with a natural disaster or public emergency and its aftermath, it should treat the individual as a temporary employee if the working relationship does not have the characteristics described above. The author has written more extensively on the distinction between independent contractors and employees, and those seeking more information should consult those publications.[20]

---

19. For the Internal Revenue Code, *see* 26 U.S.C. § 3121(d)(2). The Code does not formally define the term *employee* for purposes of determining federal income tax liability but instead provides that the usual common-law rules apply in determining the employer–employee relationship. For the Fair Labor Standards Act (FLSA), *see* Rutherford Food Corp. v. McComb, 331 U.S. 722, 726–28, 730 (1947).

20. *See Independent Contractor or Employee? The Legal Distinction and Its Consequences*, PUBLIC EMPLOYMENT LAW BULLETIN NO. 32, May 2005, www.iog.unc.edu/pubs/electronicversions/pdfs/pelb32.pdf, *and Determining Whether a Worker Is an Independent Contractor or an Employee*, POPULAR GOVERNMENT, Fall 2006, at 25–34, http://sogpubs.unc.edu/electronicversions/pg/pgfal06/article3.pdf. *See also Chapter 9, Hiring Employees versus Engaging Independent Contractors*, in RECRUITMENT AND SELECTION LAW FOR LOCAL GOVERNMENT EMPLOYERS (UNC School of Government, 2013).

Volunteers

Government employers may also accept the services of volunteers to meet the need for additional workers during an emergency response and recovery period. A government entity can draw on three types of volunteers in an emergency; different rules apply to each. The first group consists of private individuals volunteering services to a government agency, a group we will call citizen volunteers. Disasters causing extraordinary damage, such as the collapse of the World Trade Center in New York on September 11, 2001, the breaking of the levees in New Orleans during Hurricane Katrina in 2005, and the widespread, massive flooding in the New York metropolitan area following Hurricane Sandy in 2012 typically bring forth citizen volunteers on an extraordinary scale, some traveling from distant states to assist in the work of reconstruction. The second group is made up of the government unit's own employees who volunteer their services in some capacity other than the one for which they are paid. The third group consists of employees of neighboring jurisdictions with which the affected jurisdiction has a mutual aid agreement.

Citizen Volunteers

As many public employers know from their experiences with volunteer fire-fighters, emergency medical personnel, and parks and recreation coaches, governments must ensure that volunteers are not treated as employees. When government compensates volunteers for their assistance, it sometimes unwittingly and unintentionally turns those volunteers into employees, incurring tax liability for both itself and the volunteer. Citizens may volunteer for a public agency without being considered employees by the IRS or Department of Labor if they provide their services (1) freely and without coercion, direct or implied and (2) for civic, charitable, or humanitarian reasons without promise, expectation, or receipt of compensation.[21] The government agency for which the services are performed may, however, pay volunteers a nominal amount in recognition of their service or pay the expenses a volunteer incurs while performing the services.[22] There is no limit to the kinds of functions for which a person may volunteer his or her services to a government entity:

---

21. *See* 29 C.F.R. §§ 553.101(a), (c); 553.104.
22. *See* 29 C.F.R. § 553.106.

Examples of services which might be performed on a volunteer basis when so motivated include helping out in a sheltered workshop or providing personal services to the sick or the elderly in hospitals or nursing homes; assisting in a school library or cafeteria; or driving a school bus to carry a football team or band on a trip. Similarly, individuals may volunteer as firefighters or auxiliary police, or volunteer to perform such tasks as working with retarded or handicapped children or disadvantaged youth, helping in youth programs as camp counselors, soliciting contributions or participating in civic or charitable benefit programs and volunteering other services needed to carry out charitable or educational programs.[23]

This flexibility is generally helpful to government entities, but never as much as in an emergency situation, where the amount of work to be done dwarfs the number of local government employees or the emergency itself precludes large numbers of employees from being able to report to work. The U.S. Department of Labor and the federal Fourth Circuit Court of Appeals (with jurisdiction over North Carolina) have sometimes been hesitant about situations in which volunteers perform the same services as paid employees, noting,

> [I]n determining whether an employer–employee relationship exists for purposes of the FLSA, we have looked to see whether the individual seeking compensation can be said to have "displaced a bona fide applicant who desired to sell his services at prevailing rates, or . . . to be an exploited unorganized laborer, evils which the Act was designed to prevent."[24]

---

23. *See* 29 C.F.R. § 553.104(b).

24. *See* Benshoff v. City of Virginia Beach, 180 F.3d 136, 140 (4th Cir. 1999) (city firefighters who volunteered for private rescue squads were not employees when performing rescue squad activities, even where there was general oversight of the rescue squads by a city employee and the city provided the squads with loans to purchase equipment and workers' compensation coverage), *citing* Isaacson v. Penn Cmty. Servs., Inc., 450 F.2d 1306, 1310 (4th Cir. 1971) (conscientious objector, who was performing work of national importance in lieu of military service for nonprofit corporation whose purpose was public good in a position created by corporation to accommodate conscientious objectors, was volunteer not covered by the wage and overtime provisions of the FLSA).

The courts have not, however, objected to the use of volunteers for emergency services that are also provided by paid employees. The FLSA regulations expressly recognize "auxiliary police" who work alongside and at the direction of paid law enforcement officers as legitimate volunteer positions.[25] Courts have also recognized the lawfulness of having volunteer public safety personnel working with and under the direction of paid government employees.[26]

Employee Volunteers

The second group of potential volunteers consists of the government agency's own employees. In a disaster or emergency, some public employees will be deemed nonessential personnel and will not be required to report for work. Some of these "nonessential" employees may volunteer to perform their own or other jobs without pay as a community service. A public employer must proceed carefully before it accepts the services of an employee in a volunteer capacity, however compelling the need and the offer. The U.S. Department of Labor's FLSA regulations expressly exclude from the definition of *volunteer* anyone who "is otherwise employed by the same public agency to perform the same type of services as those for which the individual proposes to volunteer."[27] There are no exemptions for emergency situations from the usual rules governing employees who volunteer their services. If a government employer allows an employee to volunteer to carry out his or her regular duties, it will have to pay that employee.

The regulations do allow employees to volunteer their services without contemplation of pay when the services are not "similar or identical services" for which they are normally paid. In making such a determination, the Department of Labor will consider all the facts and circumstances in a particular case, including whether the volunteer service is closely related to

---

25. *See* 29 C.F.R. § 553.104(b).

26. *See* Cleveland v. City of Elmendorf, 388 F.3d 522, 527 (5th Cir. 2004) (so-called "nonpaid regular" police officers who received no monetary compensation from city, but who were reported as currently commissioned to the state, were volunteers and not employees, as being allowed to maintain their commissions was not a sufficiently tangible benefit to render them employees). *See also Benshoff*, 180 F.3d at 140; Todaro v. Twp. of Union, 40 F. Supp. 2d 226, 232 (D.N.J. 1999) (town's offering of eligibility for "jobs-in-blue" with private entities to special law enforcement officers who performed four hours of unpaid town duty per week did not invalidate their status as volunteers).

27. *See* 29 C.F.R. §§ 553.101(d); 553.102(a).

the actual duties performed by or responsibilities assigned to the employee.[28] The employees of a local government's planning department, for example, are unlikely to be deemed essential personnel in an emergency. Those employees could then, if they wished, volunteer to perform other work; public works duties, for example, such as assisting in clearing streets and public spaces of debris, assisting in housing and feeding citizens in government-provided shelters, or answering phone calls from citizens seeking information or assistance. Other examples of public employee volunteer service in a position that is not the "same type of service" include

> [a] city police officer who volunteers as a part-time referee in a basketball league sponsored by the city; an employee of the city parks department who serves as a volunteer city firefighter; and an office employee of a city hospital or other health care institution who volunteers to spend time with a disabled or elderly person in the same institution during off duty hours as an act of charity.[29]

As is the case with volunteers generally, employees must freely choose to volunteer their services in the different capacity and must be motivated, at least in part, by humanitarian or charitable impulses and not by any nominal fees or gifts by which the government employer expresses its thanks for the service.[30] Again, when allowing employees to volunteer, public employers must ensure that these employees will not be performing the same or similar services as those for which other employees are paid. Employees with appropriate certifications, however, may volunteer as emergency medical personnel, auxiliary or reserve police officers, or firefighters.[31]

---

28. *See* 29 C.F.R. § 103(a). *See also* 29 C.F.R. §§ 553.101(d), 553.102, 553.103.

29. *See* 29 C.F.R. § 553.103(c).

30. *See* Purdham v. Fairfax Cnty. Sch. Bd., 637 F.3d 421, 427–29, 433 (4th Cir. 2011) (the fact that school employee serving as golf coach may have been motivated, in part, by his stipend did render him an employee within meaning of the FLSA since he was not doing the same type of work required by his regular position); *Cleveland*, 388 F.3d at 527.

31. Benshoff v. City of Virginia Beach, 180 F.3d 136, 140 (4th Cir. 1999); Isaacson v. Penn Cmty. Servs., Inc., 450 F.2d 1306, 1310 (4th Cir. 1971).

Personnel Supplied through Mutual Aid Agreements

The third type of volunteer is technically not a volunteer at all. This person is employed by another government employer that, pursuant to Section 166A-19.72 of the North Carolina General Statutes, has entered into a mutual aid agreement with the affected jurisdiction to supply that jurisdiction with additional manpower (as well as supplies, equipment, facilities, and services as needed) in the event of certain circumstances. Mutual aid agreements usually provide for reciprocal services ("We'll provide you additional public safety personnel in a disaster if you agree to send us some of yours when we have emergency conditions") or for direct payment from the affected jurisdiction to the assisting jurisdiction for the cost of the services provided. In either case, the employees of the assisting jurisdiction are paid for their services by the assisting jurisdiction. If the affected jurisdiction is reimbursing the assisting jurisdiction for wages the assisting jurisdiction has paid to its employees for their work for the affected jurisdiction, then the affected jurisdiction may be able to recover some or most of those costs from FEMA (see the discussion on page 33).[32]

Retirees

A natural place for North Carolina public employers to turn for temporary assistance during an emergency is to retirees from their own or another agency or local jurisdiction. State and local government retirees hired back into the same or similar positions will not need as much training in local government generally or in the specifics of the particular department or job functions as other temporary hires. Such employees are rehired because they have proven their reliability and efficiency. Absent any emergency authorization from the North Carolina General Assembly, however, the post-retirement employment limitations on LGERS members remain in place even during a disaster and recovery period: when a retired LGERS member returns to work for an LGERS employer—whether the employer is the same employer from which the member retired or is another LGERS employer—that employee is limited in any calendar year (or during the 12-month period immediately following the effective date of retirement) to earning 50 percent of what that employee's reported compensation was in the 12-month period preceding

32. *See* U.S. Department of Homeland Security, *FEMA Disaster Assistance Policy* (DAP9523.6), www.fema.gov/pdf/government/grant/pa/9523_6.pdf.

retirement.[33] This limitation holds whether retirees return to work for the LGERS employer on a part-time, temporary, interim, or fee-for-service basis. The ability of any individual North Carolina governmental employer to hire an LGERS retiree for temporary service will therefore depend upon both how many hours and for how long the employer will need the retiree's services and how much money the retiree has already earned in that calendar year for post-retirement work for an LGERS employer.

---

33. *See* N.C. GEN. STAT. § 128-24(5)c.

Part 2

Compensation Issues: Application of the Fair Labor Standards Act to Emergency Situations

The Fair Labor Standards Act (FLSA) contains no special exemptions for compensating employees called upon to work additional hours during emergencies, to perform duties other than their regular duties, or to work under hazardous conditions. Nor does it have special rules for situations in which the employer's payroll and record-keeping systems are down or destroyed. The same FLSA rules apply in disaster situations as under normal working conditions. The following sections discuss the application of FLSA rules to situations that commonly arise under emergency operating conditions.

Overtime

When a government employer deals with an emergency, it almost always incurs significant overtime costs, particularly among public safety employees. The FLSA gives no relief in these situations. Of course, some flexibility is already provided by the section 207(k) 28-day work schedule exemption for law enforcement and firefighting employees and the fluctuating workweek exemption sometimes used for dispatchers and emergency medical service personnel. With the 207(k) schedule and the fluctuating work schedule in place, the general rule of overtime applies, even in disaster and emergencies, regardless of whether a state of emergency has been declared by the relevant federal, state, or local government authorities. For most nonexempt employees, however, the FLSA requires that they be scheduled and paid based on a 7-day workweek and that any hours physically worked in excess of 40 be paid at one and one-half times an employee's regular pay rate.

The FLSA mandates do not limit the number of hours employees may be required to work. Following a natural disaster or during a public health crisis, employees may not want to work overtime and may prefer to stay home either to tend to their families or their homes, especially if there has been extensive damage to residential areas. But employers may require their employees to work overtime (or in the case of exempt employees, in excess of their scheduled hours) and even to work significant amounts of overtime in emergencies, whether the overtime is required for clean-up activities in the jurisdiction itself or at the workplace or to cover for other employees who are absent.

Use of Comp Time in an Emergency

Public employers who credit nonexempt employees with compensatory time off (comp time) in lieu of paying cash overtime may continue to do so in emergency situations.[1] The statutory caps on the accumulation of comp time by nonexempt employees do not change under emergency conditions. Employers may allow nonexempt employees to accrue only up to 240 hours of comp time, with the exception of employees working "in a public safety activity, an emergency response activity, or a seasonal activity," who may accrue up to 480 hours.[2] Employers may only apply the 480-hour limit to employees engaged in public safety and emergency response activities as a regular part of their work. Thus, law enforcement officers, firefighters, emergency medical personnel, as well as 911 dispatchers and telecommunicators may be subject to the higher limit.[3] Employees whose regular work does not involve public safety or emergency response but who undertake such duties during the course of an emergency remain subject to the lower cap of 240 hours.[4] If disaster conditions or public emergencies require government employees to work around the clock, as they often do, the statutory limits remain in place.[5] Employers must either begin to pay out overtime hours in

---

1. Fair Labor Standards Act (FLSA) compensatory time off is credited at the statutory rate of one and one-half hours paid time off for every hour worked in excess of 40. *See* 29 U.S.C. § 207(o)(1); 29 C.F.R. § 553.22(b).

2. *See* 29 C.F.R. § 553.24(c).

3. *See* 29 C.F.R. § 553.24(c).

4. *See* 29 C.F.R. §§ 553.24(c) and (d).

5. *See* 29 U.S.C. § 207(o)(3)(A); 29 C.F.R. § 553.22(b).

cash or send employees home to use their paid time off. In a disaster, the latter is not generally an option.

Section 207(k) 28-Day Work Cycle Exemption

Under the 207(k) exemption, public employers may schedule nonexempt law enforcement officers and firefighters for work periods between 7 and 28 days[6] and are liable for time-and-one-half overtime wages only when a law enforcement officer has worked in excess of 171 hours in a 28-day period or when a firefighter has worked in excess of 212 hours in a 28-day period.[7] As is the case with public employees working in areas other than law enforcement and firefighting who follow a traditional 7-day workweek, overtime wages may be substituted with compensatory time off at a rate of one and one-half hours paid time off for every hour or fraction thereof worked in excess of 171 or 212, as applicable.[8]

Who Qualifies as a Law Enforcement Officer for 207(k) Purposes?

Only certain employees of a police department or sheriff's office may be scheduled and paid under the 207(k) exemption—namely, sworn law enforcement officers. The FLSA regulations define *law enforcement officer* as an employee who:

1. is a uniformed or plainclothes member of a body of officers,
2. has the statutory power to enforce the law,
3. has the power to arrest, and
4. has participated in a special course of law enforcement training.[9]

---

6. See 29 C.F.R. § 553.224.

7. See 29 C.F.R. §§ 553.230(a) and (b). For scheduling periods of 14 days, law enforcement officers earn overtime after 86 hours, while firefighters earn overtime after 106 hours. For 7-day work periods, the number of hours a law enforcement or firefighting employee may work before being entitled to overtime is 43 hours and 53 hours, respectively. See 29 C.F.R. § 553.230(c). Another requirement for compensating law enforcement and firefighting employees under the 207(k) exemption is that the employer make a notation in its payroll records detailing the length of the work period and its starting date and time and stating that the schedule has been adopted "pursuant to section 207(k) of the FLSA and 29 CFR Part 553." Establishing a 207(k) work period does not require approval of the U.S. Department of Labor or of employees. The work period does not have to coincide with payroll periods.

8. See 29 C.F.R. §§ 553.201(b) and 553.231.

9. *See* 29 C.F.R. § 553.211(a).

An unsworn jailer counts as a law enforcement officer for 207(k) purposes,[10] but other civilian employees of the police or sheriff's department do not.[11]

Who Qualifies as a Firefighter for 207(k) Purposes?

As is the case with law enforcement officers, only certain employees of a fire department may be scheduled and paid under the 207(k) exemption. The FLSA regulations define an employee *engaged in fire protection activities* as an employee who:

1. is trained in fire suppression;
2. has the legal authority and responsibility to engage in fire suppression;
3. is employed by a fire department; and
4. is engaged in either preventing, controlling, and extinguishing fires or responding to emergency situations where life, property, or the environment is at risk.[12]

Employees working in the fire department who do not meet this definition— so-called civilian employees—may not be scheduled and paid under the 207(k) exemption. The 207(k) exemption may be used for qualifying law enforcement and firefighter positions under all conditions, not only in emergency situations.

Dispatchers, Emergency Medical Service Personnel, and Others with Fluctuating Workweeks

The FLSA permits employers to compensate employees whose hours vary from week to week by paying them a fixed salary and paying any hours worked in excess of 40 at half their regular rate. This exception to the regular rule for overtime compensation is separate and distinct from the 207(k) exception. It is generally applied to dispatchers and emergency medical services (EMS) workers who, like their counterparts in law enforcement and firefighting, work extended shifts and different numbers of hours from week to week and are more likely to accumulate overtime hours than other local government employees.

---

10. *See* 29 C.F.R. § 553.211(f).
11. *See* 29 C.F.R. § 553.211(e).
12. *See* 29 U.S.C. § 203(y); 29 C.F.R. § 553.210(a).

To use the fluctuating workweek method, an employer must pay employees a fixed salary every week, without any variation based on the number of hours actually worked. In weeks in which employees work 36 hours, for example, they would receive the same salary as in weeks in which they work 40 or 48 hours. For any week in which an employee works in excess of 40 hours, he or she would be paid overtime at a half-time rate instead of at the time-and-one-half rate. As the regulation notes, "Payment for overtime hours at one-half such rate in addition to the salary satisfies the overtime pay requirement because such hours have already been compensated at the straight time regular rate, under the salary arrangement."[13] However, payment of a fixed salary under the fluctuating workweek exception does not turn the affected employees into exempt employees; payment of a fixed salary is only one of the requirements for a position to be exempt from FLSA rules. Positions must also satisfy one of the "duties tests" set forth in 29 C.F.R. Part 541.[14]

Under the fluctuating workweek, the employee's regular rate is calculated by dividing the salary by the total number of hours worked that week, including overtime hours. Thus, the regular rate may be different for different weeks. The U.S. Department of Labor provides a good example of how pay for fluctuating workweeks is calculated at 29 C.F.R. § 778.114(b).

Compensatory time off in lieu of cash overtime may also be used in conjunction with the fluctuating workweek, as it can with the 207(k) exemption for law enforcement officers and firefighters. The rules governing the use of the fluctuating workweek are the same in emergency situations as they are under normal working conditions.

FEMA Reimbursement of Overtime Costs

In certain instances the Federal Emergency Management Agency will reimburse state and local government entities for costs incurred in responding to a natural disaster. Sections 403, 407, and 502 of the federal

---

13. *See* 29 C.F.R. § 778.114(a).

14. For an in-depth discussion of the current duties tests, see Diane M. Juffras, *New Fair Labor Standards Act Overtime Regulations Effective August 23, 2004*, PUBLIC EMPLOYMENT LAW BULLETIN No. 31, June 2004, http://sogpubs.unc.edu/electronicversions/pdfs/pelb31.pdf.

Robert T. Stafford Disaster Relief and Emergency Assistance Act provide for federal reimbursement of some forms of emergency work labor costs of government and public not-for-profit entities.[15] This kind of reimbursement (hereinafter FEMA reimbursement) is known as a Public Assistance Program grant. It is available only after the president of the United States has made a disaster declaration applicable to the affected area and FEMA has designated the area eligible for Public Assistance Program grant funding. Only certain labor costs associated with emergency response and recovery are eligible for reimbursement under the Public Assistance Program. The cost of the straight-time wages, salaries, and benefits of a public employer's permanent employees are not eligible for reimbursement, even when the employees are performing emergency work such as implementing pre-event emergency protective measures or removing post-event debris. This is also true where the employees are performing work outside their regular duties—for example, when administrative staff assist in debris removal or in providing meals to citizens displaced by a natural disaster or when an information technology staff member who is also a volunteer fireman assists the employer's professional firefighters.[16] Permanent employees are those whose positions have been included in the employer's budget.[17] The costs of wages, salaries, and benefits for individuals sent home or told not to report due to emergency conditions are not eligible for reimbursement.[18]

---

15. *See* 42 U.S.C. §§ 5121–5207 and 44 C.F.R. §§ 204.42, 206.224, and 206.225. Qualifying public not-for-profit entities (PNPs) are generally those that provide education, medical, custodial care, emergency, utility, certain irrigation facilities, and other essential government services. Eligible PNPs are identified in 44 C.F.R. § 206.221(e).

16. *See* U.S. Department of Homeland Security, Federal Emergency Management Agency (FEMA), *Public Assistance Guide* (hereinafter FEMA *Public Assistance Guide*), at 66, www.fema.gov/pdf/government/grant/pa/paguide07.pdf.

17. *See* U.S. Department of Homeland Security, FEMA, Recovery Policy 9525.7, *Labor Costs—Emergency Work* (Nov. 16, 2006) (hereinafter FEMA Recovery Policy 9525.7), www.fema.gov/9500-series-policy-publications/95257-labor-costs-emergency-work.

18. *See* FEMA Recovery Policy 9525.7, at 42.

Reimbursement for Straight-Time and Overtime Work Repairing Damage Caused by a Disaster

FEMA classifies removing debris and implementing protective measures as "emergency work," and generally only the costs associated with overtime incurred by emergency work done by an employer's permanent employees can be reimbursed. FEMA does, however, allow reimbursement of both straight-time and overtime costs for permanent employees engaged in the repair of roads and bridges, water control facilities, buildings and equipment, and public utilities, as well as of parks and recreational facilities where those repairs are needed as a direct result of the declared disaster or emergency.[19] FEMA refers to such work as "permanent work" in contrast to "emergency work."[20]

The following FEMA press release gives examples of some of the kinds of work for which FEMA will reimburse local governments:[21]

FEMA Reimburses NYC for Emergency Overtime Response to WTC Disaster.

Release date: September 17, 2002

Release Number: 1391-153

New York, NY—The Federal Emergency Management Agency (FEMA) obligated $3.1 million today to the state of New York to help New York City with emergency overtime expenses related to the Sept. 11 attack on the World Trade Center.

The public assistance funding will reimburse the New York City Human Resources Administration's Department of Social Services (HRA) for emergency protective measures incurred from Sept. 11 to Nov. 9, 2001. This includes HRA personnel overtime expenses associated with:

- Establishing and operating the temporary NYC Office of Emergency Management (OEM) Command Center. This entailed

---

19. *See* FEMA *Public Assistance Guide*, at 66. For more detailed discussion of these categories, see 67–88.

20. *See* FEMA *Public Assistance Guide*, at 66.

21. *See* U.S. Department of Homeland Security, FEMA, *FEMA Reimburses NYC for Emergency Overtime Response to WTC Disaster,* www.fema.gov/news-release/fema-reimburses-nyc-emergency-overtime-response-wtc-disaster.

providing MIS professionals to set-up temporary information technology and telecommunications systems, as well as general relocation support.

- Retrofitting the city-managed Family Assistance Center (FAC) to ensure that all necessary information technology and tele-communications infrastructure were available to respond to the needs of affected families.
- Performing emergency visits to homebound and disabled residents within the disaster area to ensure their health and safety.

Nonexempt Overtime That Results from an Emergency

Although a government employer cannot claim reimbursement for the straight-time wage and salary costs of its permanent nonexempt employees, it can claim the cost of overtime for its permanent employees when they are performing emergency work. The reason for the distinction between straight-time and overtime compensation is that costs associated with an employer's permanent employees performing emergency work during regular working hours would be incurred regardless of whether a disaster occurred.[22] If an employer has a written comp time policy in place prior to the emergency, FEMA will base its reimbursement on that policy.[23] In most cases, this will be the equivalent of the cash overtime cost for nonexempt employees because few, if any, North Carolina local government employers award compensatory time in an amount greater than the statutory requirement of one and one-half hours for every hour worked over 40.

Overtime or Comp Time for Exempt Employees

Although a public employer may apply to FEMA for reimbursement for the overtime compensation incurred by its permanent nonexempt employees because of a disaster, the general rule is that it may not do so for permanent exempt employees. The exception is where the employer has a written policy already in place before the disaster that provides for some form of overtime or compensatory paid time off for exempt employees either generally or in emergency situations.[24] Although the policy may be limited to

---

22. *See* FEMA Recovery Policy 9525.7.
23. *See* FEMA *Public Assistance Guide*, at 45.
24. *See* FEMA *Public Assistance Guide*, at 42.

emergency situations, it may not be contingent on the availability of federal reimbursement.[25] The employer must commit to pay the overtime regardless of whether FEMA approves the cost for reimbursement. Overtime pay for exempt employees may be at a pro-rated straight-time rate (or a comp time rate of hour for hour) or at the rate of one and one-half times the pro-rated straight-time rate for every hour worked over 40 (or a comp time rate of an hour and one-half paid time off for every hour worked over 40).

Reimbursement of the Cost of Benefits

The portion of a benefits package that is dependent upon hours actually worked and that is part of an established policy is eligible for FEMA reimbursement. Such benefits may include employer contributions to the retirement system, FICA taxes and unemployment insurance taxes (as all are a standard percentage of wages), as well as the value of paid vacation and sick leave that accrues on the basis of hours worked.[26]

Reimbursement for the Cost of Outside Workers

Government employers may also seek reimbursement for the costs of independent contractors, mutual aid in accordance with an existing agreement, or temporary employees needed for emergency work.[27]

Reimbursement for the Cost of "Backfill" Staff

Under certain circumstances, FEMA will reimburse the cost of either permanent or temporary employees assigned to perform the regular duties of permanent employees who have been temporarily reassigned to emergency work covered by the Public Assistance Program. If the replacement worker is a permanent employee, only overtime compensation is reimbursable. If the replacement worker is an independent contractor or temporary employee, both straight-time and overtime costs are eligible for reimbursement. This is because the cost of an independent contractor or temporary employee, like the cost of the permanent employee's overtime, is an extra cost to the employer incurred as a direct result of the emergency conditions.[28]

---

25. *See* FEMA *Public Assistance Guide*, at 42.
26. *See* FEMA *Public Assistance Guide*, at 45.
27. *See* FEMA *Public Assistance Guide*, at 43.
28. *See* FEMA *Public Assistance Guide*, at 43.

Similarly, if the replacement worker is a permanent employee called in during what would regularly be a day off, the employer will likely incur extra costs and both regular and overtime wages may be eligible for FEMA reimbursement. In contrast, if the replacement worker is a permanent employee called in from scheduled vacation or personal leave, the leave can be rescheduled so as to eliminate the extra cost of the day's wages to the employer. Overtime compensation, however, would be eligible for reimbursement.[29]

Compensation of Salaried Employees Who Do Not Come to Work during an Emergency or Inclement Weather

Frequently, employees do not work their regular schedules during or following a natural disaster or other state of emergency. Sometimes an employee cannot get to work because travelling is dangerous or impossible, sometimes employees choose to stay home with their dependent children, and sometimes the employer itself closes for a period of time. What happens to the wages and salaries that employees would normally earn and rely upon? The rules may briefly be summarized as follows:

- Nonexempt employees who do not work do not have to be paid.
- Exempt employees do not have to be paid if they do not work for an entire workweek.
- Where the workplace remains open, an exempt employee who works for less than a full workweek may be required to use accrued paid leave for the time absent. If an exempt employee does not have accrued paid leave, then a public employer may count this as an absence for personal reasons and deduct the time lost from his or her salary.
- If conditions require an employer to close its workplace or any part of the workplace for less than a full workweek, it must pay exempt employees their full weekly salaries, although the employer may require employees to apply as much accrued paid leave as an employee has available.

The following sections discuss these rules in greater detail.

---

29. *See* FEMA *Public Assistance Guide*, at 43.

Nonexempt Employees

A simple rule applies to nonexempt employees in all circumstances: the FLSA requires employers to pay them only for the hours they have physically worked. If at any given time there is no work for an employee to perform, or if the employer decides to close on what would otherwise be a workday, a nonexempt employee is not entitled to any compensation. Most public employers, however, offer some mix of paid sick and vacation leave to their employees, both nonexempt and exempt. To alleviate the hardship that comes from not being paid an expected wage, employers may allow nonexempt employees to draw on their accrued paid leave, including accrued comp time, to turn unexpected days off caused by inclement weather into paid time.

This is true for both hourly and salaried nonexempt employees. In contrast to FLSA-exempt employees, who must be paid on a salary basis, nonexempt employees are paid on a salary basis as a matter of convenience. As with nonexempt employees paid on an hourly basis, salaried nonexempt employees must record the time they have worked on a daily basis and must be paid overtime for any hours that they physically work in excess of 40. The same rules apply to all nonexempt employees. The rules for exempt employees are different.

Exempt Employees

The general rule under the FLSA is that employers who make unlawful deductions from the salaries of their exempt employees lose the exemption and convert those employees into nonexempt employees entitled to overtime.[30] That is why employers must be careful during disaster and disaster recovery periods to follow the FLSA regulations that provide for exceptions to this rule and that will allow them to minimize the financial impact of the work days that employees miss due to emergency circumstances on their unit's budget and operations.

---

30. To be classified as FLSA-exempt, a position must not only be paid on a salary basis, but must also be paid a minimum of $455 per week, and the duties of the position must meet one of the executive, administrative, or professional duties tests. *See* 29 C.F.R. § 541.600(a). For the duties tests, *see* 29 C.F.R. Part 541.

Absent from Work for a Full Week or More or Workplace Closed for a Full Week or More

The FLSA requires employers to pay exempt employees their full salary for any week in which they have performed any work. For example, if an exempt employee works on Monday but performs no other work on any other day of the week, the employer must still pay the employee his or her full weekly salary. But where an exempt employee performs no work whatsoever in a given workweek, the employer need not pay the employee at all for that week,[31] regardless of whether the employee does not work for that full week because of illness, because of traffic or weather conditions, or because the employer tells the employee not to come to work. An employer may require employees to draw upon accrued paid leave during an absence of a full week.

Absent from Work for Less Than a Full Workweek while the Workplace Is Open

Under normal circumstances, employers require exempt employees who are absent for one or more days—full days or part days—to use accrued paid sick, vacation, or personal leave to cover the absence. The FLSA permits this because paid leave is an employer-created benefit not subject to the FLSA. When an employer pays its exempt employees their stated salary and deducts the equivalent amount of leave from their accrued leave bank, the employer satisfies the FLSA without compensating the employee for time during which no work was performed and without incurring unbudgeted salary expenses (the cost of the paid leave having already been taken into account in the employer's budget). The U.S. Department of Labor has addressed this issue several times in formal opinion letters issued by the Administrator of the Wage and Hour Division, explaining that

> [e]mployers can . . . make deductions for absences from an exempt employee's leave bank in hourly increments, so long as the employee's salary is not reduced. *If exempt employees receive their full predetermined salary, deductions from a leave bank, whether in full day increments or not, do not affect their exempt status* (emphasis added).[32]

---

31. *See* 29 C.F.R. § 602(a) for the general rule. Section 602(b) explains several exceptions to this general rule. *See also* U.S. Department of Labor Wage and Hour Opinion Letter 2009-2, dated January 14, 2009.

32. *See* Wage and Hour Opinion Letter 2009-18, dated January 16, 2009, www.dol.gov/whd/opinion/FLSA/2009/2009_01_16_18_FLSA.pdf.

Sometimes, however, an employee does not have any accrued leave upon which to draw. The general rule requiring exempt employees to be paid their full salary for any week in which they perform any work would suggest that an employer would have to pay exempt employees their full salary if they were absent for a day or two for bad weather and had no accrued leave. In both the public and private sectors, however, deductions from the salary of an exempt employee are allowed where the employee is absent for one or more full days for personal reasons other than sickness. Personal reasons other than sickness can be nearly anything. During inclement weather, personal reasons are most likely to be

- problems getting to work either because the roads are dangerous or impassable or public transportation is shut down or curtailed,
- sick dependents such as the employee's children or elderly parents needing care,
- child care issues where the regular day care provider is not operating, or
- damage to or loss of the employee's home.

Where an exempt employee is absent for personal reasons for one or more full days and part of another day, the rules governing the public and private sectors diverge. In the private sector, an employer may deduct from an employee's salary for absences for personal reasons only in full-day increments. It may not deduct pay for any partial-day absences from the employee's salary. A public sector employer, on the other hand, *may* deduct partial-day absences for personal reasons from an exempt employee's salary when that employee has no accrued paid leave, just as it may deduct pay for partial-day absences due to illness from an employee's salary when the employee has no accrued sick leave available. The public sector exception to the FLSA provides the following:

> (a) An employee of a public agency who otherwise meets the salary basis requirements of [29 C.F.R.] § 541.602 shall not be disqualified from exemption . . . on the basis that such employee is paid according to a pay system established by statute, ordinance or regulation, or by a policy or practice established pursuant to principles of public accountability, under which the employee accrues personal leave and sick leave and which requires the public agency employee's pay to be reduced or such employee to be placed on leave without pay for

absences for personal reasons or because of illness or injury of less than one work-day when accrued leave is not used by an employee because:

(1) Permission for its use has not been sought or has been sought and denied;

(2) Accrued leave has been exhausted; or

(3) The employee chooses to use leave without pay.[33]

Subsection (a)(1) of 29 C.F.R. § 541.710 allows public employers to deduct both full- and partial-day absences from the pay of exempt employees where accrued paid leave is available but the employee has asked for and the employer has denied permission to use it and the employee is absent anyway. This situation is likely to arise in an emergency where the manager or department head has designated certain employees "essential personnel" and ordered them to report to work at the same time that other "nonessential" employees are ordered or given permission to stay home. If an employee who has been deemed essential refuses to report for duty, a public employer may deduct his or her salary not only in full-day increments, as in the private sector, but in partial-day increments as well, even if that employee has accrued leave available for use. Of course, the employer may also fire the employee for insubordination.

Workplace Closed for Less Than a Full Workweek

When an emergency forces an employer to close its workplace for less than a full workweek, the employer must pay exempt employees their full weekly salaries. The federal regulation defining salary basis is explicit on this point:

An employee is not paid on a salary basis if deductions from the employee's predetermined compensation are made for absences occasioned by the employer or by the operating requirements of the business. If the employee is ready, willing and able to work, deductions may not be made for time when work is not available.[34]

Any condition that forces a public employer to close its workplace to all but essential personnel would likely keep nonessential employees at

---

33. 29 C.F.R. § 541.710.

34. See 29 C.F.R. § 601(a). *See also* U.S. Department of Labor Wage and Hour Opinion Letter 2009-2, dated January 14, 2009.

home anyhow—that is to say, employees are not going to be "ready, willing and able to work." An employer might object, perhaps not unreasonably, that it should not have to compensate employees who would not come to work if the employer stayed open. The presumption behind the regulation, however, is that employees are ready, willing, and able to work when the employer is open. When the employer closes down, exempt employees must be compensated.

Employers may, however, apply any accrued paid leave an exempt employee has to the days during which the employer is shut down:

> ... [A]n employer can substitute or reduce an exempt employee's accrued leave for the time an employee is absent from work, even if it is less than a full day and *even if the absence is directed by the employer* ... without affecting the salary basis of payment, provided that the employee still receives in payment an amount equal to the employee's guaranteed salary.[35]

But where the employee has no accrued paid leave, the employer must pay the employee his or her full salary—a situation a public employer should try to avoid:

> If an employer requires that an exempt employee work less than a full workweek, the employer must pay the employee's full salary even if: (1) the employer does not have a bona-fide benefits plan; (2) the employee has no accrued benefits in the leave bank; (3) the employee has limited accrued leave benefits, and reducing that accrued leave will result in a negative balance; or (4) the employee already has a negative balance in the accrued leave bank.[36]

35. See U.S. Department of Labor Wage and Hour Opinion Letter FLSA 2005-41, dated October 24, 2005, citing U.S. Department of Labor Wage and Hour Opinion Letters dated May 27, 1999; February 18, 1999; May 23, 1996; and April 6, 1995.

36. See U.S. Department of Labor Wage and Hour Opinion Letter FLSA 2005-41, dated October 24, 2005.

A Hypothetical Situation: The City of Paradise
Closes Its Offices during a Hurricane

Paradise, North Carolina, uses a Sunday through Saturday work schedule for FLSA purposes. A hurricane strikes on Tuesday, forcing the city to close its administrative offices Tuesday at noon. They are closed until the following Monday, although public safety personnel remain on duty on their regularly scheduled shifts throughout the week. The city has never closed its offices before, and the human resources and finance departments work furiously to run the next payroll accurately and on time. A number of questions arise and members of the team look for answers.

The payroll manager has examined some issues involving nonexempt employees. As the entire team knows, nonexempt employees, whatever department they work in and whatever their job duties are, must be paid for any hours they actually worked during that week and need not be paid at all for hours they did not work. Because the hurricane struck on Tuesday, most nonexempt employees earned at least some part of their usual weekly wages on the Sunday, Monday, or Tuesday morning of that week. The payroll manager compiles a list of those nonexempt employees who have accrued FLSA comp time balances. "Why don't we apply any accrued comp time to nonexempt employees' weekly time for the time the administrative offices were closed and an employee would have been scheduled to work except that we shut down?" he suggests to the city manager. "Some of them, at any rate, will receive all or close to all of what they would normally earn in a workweek and we will reduce our overall comp time liability." "Can we do that?" the city manager asks the human resources director. She confirms that they can. "Won't some employees grumble because they would rather wait and accumulate more comp time before using it later?" the manager asks. "Yes," the human resources director says, "they will grumble, but an employer has the right to order an employee to use comp time."[37] The manager returns to the payroll manager and compliments him. "Good thinking!" he says. "Let's do it." The team is right.

---

37. *See* Christensen v. Harris Cnty., 529 U.S. 576, 583–86 (2000).

The human resources director is thinking about issues involving exempt employees. Exempt employees will have to be paid their regular salary for the entire workweek even though most will only have worked on Monday and half of Tuesday. The only exempt employees who do not have to be paid their regular salary are those who did not perform any work whatsoever on Sunday, Monday, or Tuesday. What about exempt employees who did not perform any work on Sunday, Monday, or Tuesday because they were using approved sick or vacation leave? the human resources director wonders. After double-checking the U.S. Department of Labor's FLSA regulations, she tells the team that the city's obligation to pay employees using approved sick or vacation leave during the part of the week the offices were open is limited to the pro rata portion of their pay covered by the sick or vacation leave days. "In other words," she explains, "we need only pay them for the sick or vacation days they were using." To the quizzical looks of the others, she responds, "Sick and vacation leave is not governed by the FLSA. They did not actually physically work during the week of the hurricane. While they are paid for sick and vacation leave purposes, for FLSA purposes, they did not work at all during the week of the hurricane." "Wow," the finance director muses. "If the hurricane had struck on Sunday evening, exempt employees—other than those exempt employees working in public safety—would not have been able to perform any of their duties during that workweek and the city would not be required to pay them at all." The team is right about this, too.

The city manager has been thinking about the payroll manager's suggestion that comp time accrued by nonexempt employees be applied to the days city offices were closed and about the exempt employees who were using sick and vacation leave while the city was still open for business. He turns to the others. "Couldn't we require both nonexempt and exempt employees to use accrued vacation leave for the days the city was closed? That way most folks will receive their full wages or salary for the week and we will save some money by not having to pay exempt employees for days on which they performed no work. We'll have fewer days that people are off later in the year as well, which will improve our productivity." The human resources director and payroll manager nod. "Nothing in our personnel policies about the use of vacation leave prohibits us from doing that," the human resources director says. "I think that's a good idea. Of course, those folks who have already used up their accrued vacation leave will not be paid, but we would be in conformance with the FLSA and our policy. We'll just have to communicate clearly with our employees so that there are no misunderstandings." The team is right about this, too.

Compensation for Increased Workload and Hazardous Working Conditions Caused by a Disaster

Emergency conditions frequently require that public employees work longer hours and take on responsibilities beyond those set forth in their job descriptions. Sometimes employees must work in conditions hazardous to their own health and safety. Notwithstanding these personal sacrifices government employees are expected to make, neither the FLSA nor the Occupational Safety and Health Act of North Carolina[38] requires employers to pay employees any additional compensation for an increased workload or for hazardous working conditions. All that the law requires in an emergency situation is that nonexempt employees be paid at their regular rate for all hours worked in a given workweek up to and including hour 40 and that they be paid time-and-one-half overtime for hours over 40. If the employer uses compensatory time off, nonexempt employees must accrue one and one-half hours of paid time off for each hour worked over 40. Public safety employees compensated in accordance with the 207(k) 28-day work cycle are paid in exactly the same way in an emergency situation as they would be under normal working conditions, as are those who are compensated in accordance with the fluctuating workweek method of payment. For exempt employees, the basic rule of compensation remains the same: as salaried employees, they receive the same amount from week to week regardless of the quantity or quality of the work. In the case of a disaster, the quantity of the work will significantly increase.

---

38. Although North Carolina public employers are not subject to the federal Occupational Safety and Health Act, the Occupational Health and Safety Act of North Carolina does protect government employees and incorporates federal standards in most areas. See 29 U.S.C. §§ 651–78; 29 U.S.C. § 667; N.C. GEN. STAT. §§ 95-127(9) and (10).

Nothing prohibits an employer, of course, from being more generous to its employees than the law requires. So, for example, the Office of State Personnel has directed state agencies to compensate all employees identified as "mandatory" or indispensable during a communicable disease emergency at time and one-half their regular rate of pay for all hours worked during an emergency, not just overtime hours.[39] Similarly, the University of North Carolina's Pandemic and Communicable Disease Emergency Policy provides that non-faculty mandatory employees be paid at time and one-half during the period of the emergency regardless of their FLSA exempt status.[40] Employers may choose to compensate employees taking on additional or more hazardous duties at an out-of-class rate when the employers deem it appropriate.

Compensating Temporary Employees

Temporary employees, like their permanent counterparts, are protected by the FLSA and must be classified as either exempt or nonexempt in accordance with their job duties.[41] Temporary employees who are nonexempt must be paid overtime for all hours over 40 that they have physically worked.

The Internal Revenue Code requires employers to withhold federal income taxes from the wages of its permanent and temporary employees and generally to withhold the employee's contribution to Social Security and Medicare (FICA taxes). However, a special statutory provision exempts from Social Security or Medicare taxes temporary employees of a state or local government hired solely to assist in the organization's response to an emergency

---

39. *See* 25 N.C. ADMIN. CODE 901N .0405. This policy does not apply to local government employees subject to the State Personnel Act, whose compensation is governed by the policies of the local government board.

40. *See* The University of North Carolina Board of Governors, *UNC Pandemic and Communicable Disease Emergency Policy*, Chapter 300.2.15, adopted October 17, 2008, www.northcarolina.edu/policy/index.php?pg=vs&id=320.

41. Temporary employees are also protected by Title VII of the Civil Rights Act of 1964, the Age Discrimination in Employment Act, the Americans with Disabilities Act, and the Genetic Information Nondiscrimination Act. The only federal antidiscrimination statute that does not protect temporary employees is the Uniformed Services Employment and Reemployment Rights Act. A longer-term temporary employee who worked for an employer for over a year and for more than 1,250 hours in that year would be entitled to Family and Medical Leave Act leave for a qualifying condition.

caused by a natural disaster such as a hurricane, tornado, snowstorm, earthquake, or flood. This special exemption does not apply to an employee hired with an expectation that the position will become permanent or to permanent employees who have been reassigned from their regular duties to assist in the emergency response.[42]

The law does not require North Carolina local government employers to provide employees with any benefits other than coverage in accordance with the North Carolina Workers' Compensation Act and enrollment in one of the state's retirement systems if the employee works more than 1,000 hours a year. Under the Patient Protection and Affordable Care Act (better known as the Affordable Care Act), employers with more than 50 employees will also have to offer health insurance to any temporary or seasonal employee working 30 or more hours a week for more than three months.[43]

Other Fair Labor Standards Act and Compensation Issues
Exempt Employees Performing Nonexempt Work

As with overtime pay, the FLSA provides no exceptions to its rules governing exempt status or compensable time for disaster and emergency conditions, with one notable exception. Normally, the FLSA does not permit an employer to assign nonexempt duties to an exempt employee. However, where a dangerous condition exists or where conditions are such that an employer will have to close down if it cannot reassign some nonexempt work to exempt employees, the FLSA allows for such a reassignment of duties. The rule, which provides examples of situations that both qualify and do not qualify, is clear that this exception is only available in unusual circumstances and is to be invoked infrequently:

> (a) An exempt employee will not lose the exemption by performing work of a normally nonexempt nature because of the existence of an emergency. Thus, when emergencies arise that threaten the safety of employees, a cessation of operations or serious damage to

---

42. *See* 26 U.S.C. § 3121(b)(7)(iii); 42 U.S.C. § 418(c)(6). See also Internal Revenue Service, *Federal-State Reference Guide*, Publication 963, Rev. 11-2013, at 5-7; 6-10, Question 6, www.irs.gov/pub/irs-pdf/p963.pdf.

43. *See* 26 U.S.C. § 4980H(c)(2).

the employer's property, any work performed in an effort to prevent such results is considered exempt work.

(b) An "emergency" does not include occurrences that are not beyond control or for which the employer can reasonably provide in the normal course of business. Emergencies generally occur only rarely, and are events that the employer cannot reasonably anticipate.

(c) The following examples illustrate the distinction between emergency work considered exempt work and routine work that is not exempt work:

(1) A mine superintendent who pitches in after an explosion and digs out workers who are trapped in the mine is still a bona fide executive.

(2) Assisting nonexempt employees with their work during periods of heavy workload or to handle rush orders is not exempt work.

(3) Replacing a nonexempt employee during the first day or partial day of an illness may be considered exempt emergency work depending on factors such as the size of the establishment and of the executive's department, the nature of the industry, the consequences that would flow from the failure to replace the ailing employee immediately, and the feasibility of filling the employee's place promptly.

(4) Regular repair and cleaning of equipment is not emergency work, even when necessary to prevent fire or explosion; however, repairing equipment may be emergency work if the breakdown of or damage to the equipment was caused by accident or carelessness that the employer could not reasonably anticipate.[44]

Thus, a local government employer responding to an emergency is not likely to lose the exemption when an exempt employee

- performs first responder work,
- performs maintenance or repair work on vehicles or equipment needed in the response effort,

44. *See* 29 C.F.R. § 541.706.

- drives vehicles or equipment needed to clear roads of debris or other spaces of material posing an imminent threat of danger,
- performs nonexempt duties in a water treatment plant, or
- orders supplies or makes payments necessary for the immediate recovery effort.

Similarly, in a public health emergency such as the spread of pandemic flu or the release of chemical or biological weapons by a terrorist group, government employers are not likely to forfeit an employee's exempt status when that employee performs the work of an emergency medical technician, paramedic, licensed practical nurse, or health aide when a patient needs immediate care. These are examples only and not an exclusive list.

On-Call Time

The rules governing the compensability of on-call time do not change in emergencies. On-call time (sometimes referred to as standby time) is time that employees spend off the employer's premises during which they can pursue their own interests but still remain available to be called back into work on short notice if the need arises. The general rule is that where the employer requires an employee to be on call while remaining at the workplace or while so close to the workplace as to make the distinction immaterial (within a five-minute or five-mile radius, for example), the time spent on call is equivalent to time worked and is compensable at the employee's regular rate (or at the employee's overtime rate where applicable).[45] Where the employer does not require an on-call employee to remain at the workplace or very nearby, but merely to be reachable and available to return to work, the time during which the employee is subject to recall does not have to be paid.[46] An employer may compensate on-call employees at their regular rate or with a nominal fee, if it so chooses.

An employer who does not usually require on-call employees to remain nearby may do so in an emergency. Thus, in anticipation of an increased need for first responders or crews to clear the roads, for example, a government employer might direct law enforcement officers, firefighters or paramedics, or heavy equipment operators who are not scheduled to work to remain at their worksites without actually assigning them work to perform. This

---

45. *See* 29 C.F.R. § 785.17.
46. *See* 29 C.F.R. §§ 551.431 and 785.17.

time would count as time worked and would be compensable. Conversely, an employee placed on call because of an impending storm might think it likely he or she will be called in and prefer to remain at the worksite rather than brave treacherous conditions to return. In that case the time spent on call is not compensable, as the employer has not directed the employee to stay on-site.[47]

Increased Travel Time

Sometimes emergency conditions render the employer's worksite unusable and require an employer to relocate its base of operations. For some employees, the commute will then be longer; for others, shorter. Employees themselves may be displaced from their homes after natural disasters and face longer commutes to work from their temporary locations. In neither case is an employee entitled to compensation for the increased travel time from "home" to the worksite.[48]

Record Keeping
Recording Hours Worked Remotely

Some employers allow nonexempt employees to telecommute in the normal course of business and already have a method of tracking the hours worked remotely. Employers who do not may think record keeping for employees working off-site during a storm and its aftermath, or even during a public health emergency, would be relatively simple. Indeed, during the short term, instructing employees to email in a digitally signed timesheet to their supervisors or to payroll might well suffice. But what happens when electricity or email is interrupted? Or landline or cell phone service? Or if flooding or wind damage takes out the organization's information technology infrastructure?

Employers should develop procedures for recording the work time of nonexempt employees in two situations: one where the technology upon which we increasingly depend is available and one where it is not (without this technology many employees will be unable to work even from home). Determining the best procedure will depend on the resources of the

---

47. *See* 29 C.F.R. §§ 551.431(a)(2) and 785.17.
48. *See* 29 C.F.R. § 785.35.

organization itself and its employees (not all employees, for example, will have home computers), as well as the organization's size and reporting structure. The key point for employers to remember is that there is no emergency-situation exemption from maintaining accurate records. While enforcement agencies such as the U.S. Department of Labor and the Internal Revenue Service might grant employers filing or record-keeping extensions, these employers will ultimately need a way to determine the number of hours worked by each employee. Even with a plan in place, employers may have to rely on the records or recollections of the employees themselves. Without a plan, they will certainly be forced to do so. Remember that under normal conditions, where an employer has no records to prove otherwise, the U.S. Department of Labor relies on employee representations of the number of hours of overtime worked. The same will hold true in emergency situations.

Destruction of Payroll-Related Records and Software Applications

Running payroll requires more than simply having a record of the number of hours worked by each employee. Information about employee income tax withholding and bank account and routing numbers are also key components of paying employees. Much of this information is likely to be in an employer's computer system and available if the system itself is up and running. But natural disasters may render that information unavailable. Employers should back up all information needed to run payroll, including Social Security numbers for tax reporting purposes, bank account numbers, and bank routing information. Backup should be off-site and easily accessible. The increasing availability of secure cloud storage makes this much easier to accomplish than it has been previously.

The Fair Labor Standards Act requires employers to pay employees at regular intervals. In the event of a disaster, employers will still be legally required to pay employees at the next regular payroll—and under emergency conditions, employees may need to be paid at the next regular payroll to get by. In addition to backing up the information needed to run payroll, employers should ensure that they have network redundancy and backup access to the software needed to run payroll and transfer wages and taxes to the applicable accounts.

Part 3

Issues Specific to Public Health Emergencies

Public health emergencies require special planning. An outbreak of pandemic flu or SARS (severe acute respiratory syndrome) or a terrorist attack that disperses biochemical or chemical agents such as anthrax, ricin, or smallpox presents a unique set of issues that employers must address in addition to those arising from natural disasters more generally. Employees may be contagious or sick. Public health authorities may issue quarantine or isolation orders affecting employees' ability to come to work. Employers may want to monitor the health of their employees in the workplace. Public health employees may be fearful of working with a contagious or sick population. Employers may want to ask employees to undergo vaccination against the threat. Requests for Family and Medical Leave Act (FMLA) leave and for Americans with Disabilities Act (ADA) accommodations are likely to increase. Public health employees will have questions about workers' compensation coverage. The questions are complicated, as are some of the answers. Employers would be wise to address these issues before the onset of a public health crisis.

Mandatory Vaccinations

Employers around the country have long encouraged their employees to receive annual flu vaccinations. Some health care employers have made receiving an annual flu vaccination a condition of employment for those who have patient contact. The strains of influenza that have circulated in recent years have not been especially deadly nor more transmissible than usual. Thus, employers outside the health care setting have not had to face the question of whether to require vaccination of all employees. But what if SARS were to break out again and spread more widely than it did in 2003? Or if viruses that only infect human beings through direct contact with animals mutate to make human-to-human transmission possible?[1] In December 2002 the federal government announced a voluntary (and ultimately unsuccessful) national smallpox vaccination program to strengthen the ability of the United States to protect itself from a smallpox attack by terrorists or hostile governments. What if there were some indication that forces hostile to the country were likely to weaponize smallpox? Could North Carolina public employers require health care workers, first responders, and other employees to be vaccinated against such threats as a condition of their continued employment?

Public Employers May Require Vaccination

Nothing prohibits a North Carolina public employer from requiring some or all of its employees to be vaccinated against particular illnesses. Currently North Carolina's laws requiring vaccination of adults are limited in scope, applying primarily to flu vaccination of workers in adult care and nursing homes.[2] Nothing in the U.S. Constitution, the North Carolina Constitution, or federal statutory law would prohibit the General Assembly from adopting a law requiring vaccination, however. Indeed, the state already requires vaccinations for children. In North Carolina "[e]very child present in this state" must be immunized against certain diseases.[3] A state rule specifies which vaccines are required, how many doses of each are required, and when a child should receive each dose.[4] For example, the rule requires four

---

1. *See, e.g.,* Centers for Disease Control and Prevention (CDC), *Public Health Threat of Highly Pathogenic Avian Influenza A (H5N1) Virus,* www.cdc.gov/flu/avianflu/h5n1-threat.htm.

2. See N.C. GEN. STAT. (hereinafter G.S.) §§ 131D-9 and 131E-113, respectively. North Carolina Secretary of Health and Human Services Wos has adopted a policy requiring employees in her department to be vaccinated against flu. North Carolina's communicable disease laws also provide for mandatory vaccination against certain diseases, such as Hepatitis B, in certain circumstances.

3. *See* G.S. 130A-152.

4. *See* 10A N.C. ADMIN. CODE (hereinafter N.C.A.C.) 41A .0401.

doses of polio vaccine: two before the child reaches the age of five months, a third before the child reaches nineteen months, and a fourth before the child enters school. Children must be immunized according to this schedule, which appears in the North Carolina Administrative Code.[5] Thus, in the event of another outbreak of the SARS virus, for example, a North Carolina county could order all employees to be vaccinated against SARS (assuming such a vaccine was available) and to show evidence of vaccination. With limited exceptions discussed below, the county could terminate or put on leave without pay any employee who refused to do so.[6]

Imagine the following scenario:

As scientists had feared, a mutation in the H5N1 avian flu (bird flu) virus has made the virus capable of human-to-human transmission. Previously it could be contracted by a human being only through direct contact with an infected bird. Now the bird flu can travel from person to person just like the other influenza viruses that occur every year. But unlike other flu viruses, infection with H5N1 has a 60 percent mortality rate.[7] When the first cases appear in North Carolina, the manager of Paradise County, county health director, emergency medical services director, and emergency management director meet to discuss the likely impact. The health director says a vaccine is readily available and the health department will soon offer vaccination clinics to residents. The group discusses encouraging all county

---

5. Immunization rules are adopted by the North Carolina Commission for Public Health. G.S. 130A-152(c) (authorizing the commission to adopt rules implementing an immunization program); see also G.S. 130A-29 (establishing the commission and giving it rule-making authority). Such rules are effective statewide and have the force of law.

6. For employees of the county and district health departments and county social services and emergency management departments protected by the State Personnel Act, the due process procedures set forth in the Administrative Code would have to be followed.

7. See World Health Organization, *Influenza: Facts: H5N1 Influenza*, www.who.int/influenza/human_animal_interface/avian_influenza/h5n1_research/faqs/en/index.html. Some researchers question whether the World Health Organization's 60 percent mortality rate is too high. See Declan Butler, *Death-Rate Row Blurs Mutant Flu Debate*, NATURE, Feb. 16, 2012, at 289, www.nature.com/news/death-rate-row-blurs-mutant-flu-debate-1.10022; Helen Branswell, *Dread Reckoning: H5N1 Bird Flu May Be Less Deadly to Humans Than Previously Thought—or Not*, SCIENTIFIC AMERICAN, Feb. 14, 2012, www.scientificamerican.com/article.cfm?id=h5n1-bird-flu-case-fatality-calculations.

employees to receive a vaccination and scheduling employee-only vaccination clinics during paid working hours to make it as easy as possible for employees to become immunized.

The health director says she plans to require all health department employees to be immunized. Any employees who fail to be vaccinated through the health department or to provide evidence that they have received a vaccination from their private provider will be suspended without pay until they do so. The health department's resources, she explains, will be heavily taxed in the event of a flu outbreak. It is imperative, she says, that the health department (1) prevent infected, but not yet symptomatic, employees from spreading the virus to the public it serves and (2) ensure it has sufficient staff available and able to provide flu-related and other important services. She sees no way around mandatory vaccination.

The group consults with the county commissioners about its plan to encourage voluntary immunization and to offer employee-only vaccination clinics. One of the commissioners proposes that the county require vaccination of all employees under the supervision of the county manager and that they ask the social services director to order his employees to be vaccinated. "Do you see any problems with this?" one of the other commissioners asks the county attorney. "No, I do not," says the county attorney, "provided that we give exemptions for employees who have medical conditions that prevent them from getting a bird flu shot or who object to vaccinations on religious grounds."

The county attorney is right.

Medical Exemptions from a Mandatory Vaccination Requirement

An employee with a disability that prevents him or her from receiving a particular vaccination must be given an exemption when an employer makes that vaccination mandatory. The ADA requires employers to provide a reasonable accommodation for the known physical limitations of employees unless doing so would impose an undue hardship on the employer's operations. The obligation to make reasonable accommodation is a form of non-discrimination.[8] The Equal Employment Opportunity Commission

---

8. *See* 29 C.F.R. Part 1630, App. 1630.9.

(EEOC) has said that it considers it a reasonable accommodation to exempt an employee from a mandatory vaccination requirement when the employee has an ADA-covered disability that prevents the taking of that vaccine. For this reason the EEOC advises employers to consider simply encouraging employees to get vaccinated rather than making vaccination mandatory.[9] Although the EEOC advises against mandatory policies to reduce the likelihood of a reasonable accommodation claim, an employer may still require vaccination for some or all of its employees.

North Carolina has few laws and regulations governing the vaccination of adults. It does, however, provide for mandatory vaccination of children. As in the case of employees with disabilities that have contraindications for vaccination, the child immunization law provides exemptions from its requirements for children with medical conditions for which vaccination generally or a particular vaccination is contraindicated.[10]

An employer is not required to accommodate an employee's disability where to do so would cause an undue hardship. An undue hardship is "a significant difficulty or expense" when considered in light of the (1) nature and net cost of the accommodation needed; (2) overall financial resources of the employer; (3) number of persons working for the employer; (4) type of operation of the employer, including the composition, structure, and functions of the workforce and the geographic separateness and administrative or fiscal relationship of the specific worksite to the employer; and (5) impact of the accommodation upon the employer's operation, including the impact on the ability of other employees to perform their duties and on the employer's ability to conduct business.[11]

What is an undue hardship in this context? In light of the factors set forth by the EEOC, the circumstances would have to be quite extraordinary to justify either terminating an employee who cannot safely be vaccinated or suspending him or her without pay. In most cases the employee should be allowed to continue to work or to take paid leave. Where the goal of mandatory vaccination is to ensure that government operations can continue without excessive absenteeism, the exemption of a single employee is unlikely

9. *See* Equal Employment Opportunity Commission (EEOC), *Pandemic Preparedness in the Workplace and the Americans with Disabilities Act* (hereinafter *Pandemic Preparedness in the Workplace*), Question 8, www.eeoc.gov/facts/pandemic_flu.html.
10. *See* G.S. 130A-156.
11. *See* 29 C.F.R. § 1630.2(p).

to cause undue hardship: employees who have been vaccinated will not contract the illness from the exempted employee. If the exempted employee is in a position that requires contact with the public, then the same common-sense, everyday rules should apply: employees showing obvious signs of illness should be sent home until they are asymptomatic and, if they are diagnosed with infectious disease, until they are no longer contagious. This is consistent with the EEOC's position as set forth in its technical guidance on pandemic preparedness. There, the agency says that an employer may lawfully exclude a person with a disability from employment in a pandemic only when it can demonstrate that the employee poses a direct threat.[12] As part of an accommodation, an employer could also require an employee who is exempt from a vaccination requirement for medical reasons to wear a mask, provided that the medical condition does not preclude it.

Religious Exemptions from a Mandatory Vaccination Requirement

Similarly, under Title VII of the Civil Rights Act of 1964, once an employer receives notice that an employee's sincerely held religious belief prevents that employee from taking a vaccine, the employer must provide a reasonable accommodation unless doing so would pose an undue hardship. This is again consistent with the approach taken under child vaccination laws. Section 130A-157 of the North Carolina General Statutes (hereinafter G.S.), for example, provides an exemption from vaccination for children whose parents or guardians have bona fide religious objections to immunization.

A claim of religious discrimination under Title VII can be proven using either direct evidence or the *McDonnell Douglas* burden-shifting framework. To establish a prima facie case that an employer has failed to accommodate an employee's religious objections to vaccination, the employee must establish that: (1) the employee has a bona fide religious belief that conflicts with an employment requirement (here, the requirement that the employee be vaccinated against a disease); (2) the employee informed the employer of this belief; and (3) the employee was disciplined for failure to comply with the requirement.[13]

---

12. *See Pandemic Preparedness in the Workplace,* Question 14, www.eeoc.gov/facts/pandemic_flu.html.

13. *See, e.g.,* EEOC. v. Firestone Fibers & Textiles Co., 515 F.3d 307, 312 (4th Cir. 2008) (employee established prima facie case); Chalmers v. Tulon Co. of Richmond, 101 F.3d 1012, 1019 (4th Cir. 1996) (employee satisfied first and third prongs of test but failed

If the employee establishes a prima facie case, the burden then shifts to the employer to show that it cannot reasonably accommodate the employee's religious beliefs without undue hardship. This is a two-pronged inquiry. To satisfy its burden, the employer must demonstrate either that (1) it provided the plaintiff with a reasonable accommodation for the plaintiff's religious observances or (2) such accommodation was not provided because it would have caused an undue hardship.[14] Under Title VII, an undue hardship is "more than de minimis cost" to the operation of the employer's business, a lower standard than under the ADA.[15]

It is difficult for an employer to determine whether an employee's sincerely held religious beliefs preclude receiving vaccinations.[16] An employer may take at face value the assertions of employees who are practicing Christian Scientists that their religion prohibits them from immunization, as it is well known that Christian Science generally rejects traditional medicine. But what about claims that other religious beliefs reject the practice of vaccinations, beliefs with which the employer may not be familiar? Employers will likely have to accommodate them. Title VII defines religion broadly to include "all aspects of religious observance and practice, as well as belief . . . ."[17] In its regulations implementing Title VII, the EEOC says that "religious practices . . . include moral or ethical beliefs as to what is right and wrong which are sincerely held with the strength of traditional religious views."[18] In order to be protected by Title VII, says the U.S. Supreme Court, religious beliefs need not be "acceptable, logical, consistent, or comprehensible to

---

to notify employer of her beliefs). For a discussion of the *McDonnell Douglas* burden-shifting framework, see DIANE M. JUFFRAS, RECRUITMENT AND SELECTION LAW FOR LOCAL GOVERNMENT EMPLOYERS 38–42 (UNC School of Government, 2013).

14. *See Firestone Fibers & Textiles Co.*, 515 F.3d at 312 (employer reasonably accommodated employee's need to attend religious services even if it did not totally accommodate employee's request); EEOC v. Thompson Contracting, Grading, Paving, & Utilities, Inc., 793 F. Supp. 2d 738, 744 (E.D.N.C. 2011), *aff'd*, 499 F. App'x 275 (4th Cir. 2012).

15. *See Firestone Fibers & Textiles Co.*, 515 F.3d at 312.

16. The U.S. Supreme Court noted that determining whether a belief is religious is "more often than not a difficult and delicate task" and one to which the courts are ill-suited. *See* Thomas v. Review Bd. of Ind. Emp't Sec. Div., 450 U.S. 707, 714 (1981).

17. *See* 42 U.S.C. § 2000e(j).

18. *See* 29 C.F.R. § 1605.1.

others . . . ."[19] It is fair to say that Title VII does not leave an employer much opportunity to challenge the religious nature of an employee's professed beliefs. This is so despite that fact that the law does not (1) recognize a non-religious "philosophical" or "personal" exemption to immunizations[20] nor (2) require an employee's "preference" or "desire" to exercise his or her religion in a particular manner to be accommodated where a particular practice or observance is not required by the religion.[21]

Would a public employee's refusal to be vaccinated against a pandemic flu or another dangerous disease based on a religious objection constitute an undue hardship for a public employer? The answer is unclear. Research has

---

19. *See Thomas*, 450 U.S. at 714; EEOC v. Union Independiente de la Autoridad de Acueductos y Alcantarillados de Puerto Rico, 279 F.3d 49, 56 (1st Cir. 2002). One of the better known cases about religious beliefs unfamiliar to others involved a Costco cashier who claimed that she belonged to the Church of Body Modification, which involved the wearing of facial piercings. The court found that accommodating the plaintiff would be an undue hardship to Costco, which had a policy prohibiting facial jewelry worn by cashiers, and noted that this finding meant the court did not have to tackle the "thorny" question of the sincerity of the plaintiff's beliefs. *See* Cloutier v. Costco Wholesale Corp., 390 F.3d 126, 132 (1st Cir. 2004).

20. *Cf.* 10A N.C.A.C. 41A .0403 ("[T]here is no exception to these requirements for the case of a personal belief or philosophy of a parent or guardian not founded upon a religious belief."), one of the regulations issued under North Carolina's child vaccination laws.

21. *See* Dachman v. Shalala, 9 F. App'x 186, 192 (4th Cir. 2001) (employee not entitled to reduced Friday work schedule where her religious beliefs prohibited working on the Sabbath but did not mandate that all preparation for the Sabbath take place on Friday); Bush v. Regis Corp., 257 F. App'x 219, 221 (11th Cir. 2007) (employer not liable for religious discrimination where assignment to Sunday shift was objectionable because plaintiff wished to perform field service on that day, but field service was not required to be performed on Sundays). See also Reed v. Great Lakes Cos., Inc., 330 F.3d 931, 935 (7th Cir. 2003); *Union Independiente de la Autoridad de Acueductos y Alcantarillados de Puerto Rico*, 279 F.3d at 56 (genuine issue of material fact as to whether Seventh Day Adventist's objection to union membership was product of sincerely held religious belief); Seshadri v. Kasraian, 130 F.3d 798, 800–01 (7th Cir. 1997) (employee refused to identify religion that required scrupulous honesty); Vetter v. Farmland Indus., Inc., 120 F.3d 749, 752–53 (8th Cir. 1997) (whether employer's requirement that employee live in particular town interfered with observance or practice of employee's religion or whether he chose to live elsewhere because of purely personal preference was jury question); *but see* Brown v. Pena, 441 F. Supp. 1382 (S.D. Fla. 1977) (even though the ancient Egyptians worshipped cats, a belief in the deeply spiritual effects of eating Kozy Kitten People/Cat Food is not a religious belief, and the plaintiff's consumption of the cat food was therefore not a religious observance).

not revealed a single case in which this issue is addressed. And even if it did, how such a determination would benefit an employer remains uncertain. A court will never force an employee to receive a vaccination as a condition of employment.[22] All such a determination would do is absolve the employer of liability in the event it terminated the employee. But in a disaster, a public employer is likely to need all of its employees and the better course would be to allow the employee to continue to work without receiving the vaccination, if the particular job duties make that feasible, or to send the employee home, with or without pay in accordance with the employer's policies.

Workers' Compensation Coverage for Adverse Reactions to a Vaccine

"What about workers' comp?" asks a third Paradise County commissioner. "What if employees have an adverse reaction to the vaccine that we have forced them to take?" The county attorney has researched this issue in anticipation of just such a question. "If we adopt a requirement that all or only specified groups of employees must be vaccinated, those employees will be covered by our workers' compensation policy for any adverse medical reactions that they may have to the vaccine. Workers' compensation will cover injuries from vaccination even if we do not require the vaccination, but merely encourage it by allowing employees to use work time to be vaccinated, either on-site or off. This issue has come up repeatedly over the past few years and the answer is clear."

The attorney is right.

In North Carolina, as in most states, workers' compensation does not require a showing of negligence or fault. Generally, the North Carolina Workers' Compensation Act (the Act) covers medical expenses incurred by the employee as a result of a work-related injury, lost wages due to the employee's temporary or permanent inability to work, and, in the case of death, compensation to the employee's designated beneficiaries for a limited period of time. To recover lost wages and medical expenses, employees need show only that they suffered a covered illness or injury and that the illness or injury occurred

---

22. A court could order a person (who happens to be an employee) to be vaccinated if vaccination is a required communicable disease control measure under North Carolina's public health laws. See G.S. 130A-144(f), 130A-18, and 130A-25.

as a result of their employment.[23] In order to recover benefits under the Act, an employee must prove that (1) he or she suffered an injury by accident; (2) the injury occurred while the employee was engaged in some activity or duty he or she is authorized to undertake and which is in furtherance of the employer's business; and (3) the injury was sustained in the course of the employment—that is, the injury occurred while the employee was working, in a place the employee might reasonably be expected to be while working, and the activity in which the employee was engaged when injured was related at least incidentally to his or her employment.[24]

Two North Carolina cases are relevant to the issue of workers' compensation coverage for required vaccinations. The first, a 1957 North Carolina Supreme Court decision, held that an injury sustained by an employee while taking a medical test or exam required for continued employment was an accident arising out of and in the course of employment within the meaning of the North Carolina Workers' Compensation Act.[25] In that case, a dairy farm employee was injured while a nurse drew blood from her arm as part of a Wassermann Test for syphilis, which the county health department required of all employees who bottled milk. At the time of the court's decision, the question of whether an injury from vaccination or immunization required by an employer was compensable under the workers' compensation act had not yet arisen, but the court noted,

> Cases where employees have been injured as a result of vaccination or inoculation present similar legal problems. There are only a few of them and all which have been called to our attention have been considered. What they hold is, that the right to compensation is determined by the construction placed on the phrase, "Arising out of and in the course of employment," and its application to the facts in each particular case. Compensation was allowed in those cases

23. In exchange for not requiring that an employee show fault, the North Carolina Workers' Compensation Act does not allow the injured employee to recover damages for pain and suffering or to seek punitive damages against the employer for its alleged fault.

24. *See, e.g.,* Cole v. Guilford Cnty., 259 N.C. 724 (1963); Taylor v. Twin City Club, 260 N.C. 435, 438–39 (1963); Fulcher by Wall v. Willard's Cab Co., 132 N.C. App. 74, 81–82 (1999); Sisk v. Tar Heel Capital Corp., 166 N.C. App. 631, 635 (2004).

25. See King v. Arthur, 245 N.C. 599 (1957).

where the vaccination or inoculation was performed for the benefit of the employer, and had been required by him.[26]

In 2008, in the case *Kai-Ling Fu v. UNC Chapel Hill*, the direct question of whether the side effects from an employer-mandated vaccination were covered under the North Carolina Workers' Compensation Act came before the North Carolina Court of Appeals for the first time.[27] In this case a university lab researcher was required to be vaccinated against the equine encephalitis virus that was used to study the AIDS virus. The vaccine caused the researcher to suffer upper respiratory symptoms, fever, headache, and extreme fatigue, which did not subside for several months. The court of appeals did not analyze the issue the same way as had the North Carolina Supreme Court in its 1957 decision. Rather than looking at the complications as resulting from an "accident," the court of appeals considered whether the researcher's side effects were an occupational disease within the meaning of the workers' compensation act.

On its face, "injury" might not seem to include "illness," and the Act's definition of *injury* explicitly excludes "disease in any form, except where it results naturally and unavoidably from . . . [an] accident."[28] Nonetheless, some illnesses, usually referred to as "occupational diseases," are covered. G.S. 97-52 provides that disability or death resulting from an occupational disease is to be treated as an injury by accident within the meaning of the Act and is therefore compensable. G.S. 97-53, in turn, lists specific diseases and conditions that may be recognized as occupational diseases. G.S. 97-53 also contains a "catch-all" subsection (13), which defines as a covered occupational disease any disease

> which is proven to be due to causes and conditions which are characteristic of and peculiar to a particular trade, occupation, or employment, but excluding all ordinary diseases of life to which the general public is equally exposed outside of the employment.

Thus, under the North Carolina Workers' Compensation Act, to establish the existence of a compensable occupational disease an employee must show that (1) the disease is characteristic of individuals engaged in the particular

26. *See King*, 245 N.C. at 602 (citations omitted) (citing cases from other states).
27. *See* 188 N.C. App. 610 (2008).
28. *See* G.S. 97-2(6).

trade or occupation in which the claimant is engaged, (2) the disease is not an ordinary disease of life to which the public generally is exposed to the same extent as those engaged in that particular trade or occupation, and (3) there is a causal relationship between the disease and the claimant's employment.[29]

In the *Fu* case, the court of appeals held that the researcher's claim met the statutory requirements. First, her illness was characteristic of persons working in labs using the equine encephalitis virus, as the vaccine was approved only for persons doing research with it and a percentage of researchers receiving the vaccine suffered side effects. Second, her illness met the requirement that it was not an ordinary disease of life to which the general public is equally exposed because the Food and Drug Administration had not approved the vaccine for public use. Finally, the trial court had given greater weight to the testimony of those physicians who found the researcher's symptoms were caused by the vaccine than it had to those who thought the symptoms were caused by something else.[30]

It isn't clear that the reasoning in the *Fu* case, which involved a rare vaccine against a relatively uncommon virus, should be applied to the question of whether workers' compensation will cover injuries arising out of employer-mandated vaccinations required as a result of a regional, national, or global pandemic. If a majority of the population is getting or being encouraged to get a flu vaccine, vaccine-related injuries will likely not meet the second requirement necessary to show they are occupational diseases—namely, that injuries or side effects from a flu vaccine are not ordinary diseases of life to which the public generally is equally exposed.[31] In a pandemic or even a less widespread outbreak in which vaccination is common, the general public will be equally exposed to the possibility of complications from the vaccine.[32]

---

29. *See* Rutledge v. Tultex Corp., 308 N.C. 85, 93–94 (1983); Johnson v. City of Winston-Salem, 188 N.C. App. 383, *aff'd*, 362 N.C. 676 (2008).

30. *See Fu*, 188 N.C. App. at 613–15.

31. *See* G.S. 97-57(13); *Fu*, 188 N.C. App. at 610, 613; *Rutledge*, 308 N.C. at 93.

32. See *Booker v. Duke Medical Center*, 297 N.C. 458, 470–72 (1979), where the North Carolina Supreme Court found that serum hepatitis, when contracted by a hospital laboratory technician, was an occupational disease and not a noncompensable "ordinary disease of life" because there was a recognizable link between the nature of the job and an increased risk of contracting the disease. The court nevertheless recognized that G.S. 97-53(13) precludes coverage for those ordinary diseases of life to which the general public is equally exposed. *See also* Hassell v. Onslow Cnty. Bd. of Educ., 182 N.C. App. 1, 11–12 (2007), *aff'd as modified*, 362 N.C. 299 (2008) (sixth

Many other states have recognized complications from vaccination as a compensable workers' compensation injury when an employer requires vaccination. They have done so using the same analysis as did the North Carolina Supreme Court in the 1957 case involving the injury suffered by a dairy farm worker during a blood draw. In that case, and in the analogous cases from other states, the court framed the issue as one of whether the complications from vaccination were an injury by accident that occurred during or in the course of employment rather than whether the complications were an occupational disease. In a relatively recent Massachusetts case, the flu shot given by a hospital to one of its employees caused the employee to become blind. Massachusetts law, like North Carolina law, requires that a compensable injury arise out of and in the course of employment. The Massachusetts court found that the employee's injuries were compensable (1) because the vaccine was administered by the hospital itself during the employee's lunch hour and the hospital had encouraged her to be vaccinated and (2) because the employee was a health care worker having direct contact with patients, the employee's vaccination furthered her employer's interests by reducing the likelihood that hospital employees would spread the virus.[33]

In another flu shot case, a Delaware court found that the injury to an employee who developed Guillain-Barre Syndrome and chronic polyneuropathy from a flu shot funded by her employer and administered at the workplace by the employer's health service arose out of her employment and was covered by that state's workers' compensation law, even though the employer offered the shots merely as a convenience for its employees.[34] Courts in other states where the requirement is that the injury must have arisen in the course of and out of employment have made similar findings.[35] Still other cases have found vaccinations against conditions other than flu to have arisen in the course of or out of employment when the vaccination has been required by the employer.[36]

---

grade teacher did not face stresses and challenges unlike those faced by the general public).

33. *See* Case of Hicks, 820 N.E.2d 826, 833–36 (Mass. App. Ct. 2005).

34. *See* E.I. DuPont De Nemours & Co. v. Faupel, 859 A.2d 1042, 1052–54 (Del. Super. Ct. 2004).

35. *See* Monette v. Manatee Mem'l Hosp., 579 So. 2d 195 (Fla. Dist. Ct. App. 1991); City of Austin v. Smith, 579 S.W.2d 84 (Tex. App. 1979).

36. See Washington Hosp. Ctr. v. Dist. of Columbia Dep't of Emp't Servs., 821 A.2d 898 (D.C. 2003) (complications from MMR vaccine covered by workers' compensation

"Does our workers' compensation liability depend in any way on whether the county has required the vaccine or simply offered it and encouraged employees to get it?" asks another Paradise county commissioner.

The voluntary nature of an employee's decision to be vaccinated against a disease would not affect whether injuries resulting from the vaccine are covered by the North Carolina Workers' Compensation Act. It is a common misconception that "voluntary acts" undertaken in the course of employment are automatically exempt from coverage under the Act. This is simply not the case. The standard for compensability has always been whether an injury or illness arises out of the injured person's employment and occurs in the course of that employment rather than the voluntariness of the employee's action. This misconception is probably an overgeneralization of the holdings of several North Carolina appellate court cases finding injuries sustained at employment-related recreational events were not compensable in the absence of any evidence that attendance was not voluntary and that the employer sponsored, financed, or derived some benefit from the event.[37]

---

where local regulations required employee to receive vaccine before beginning work in hospital); Petit v. Scipio Volunteer Fire Dist., N.Y.S.2d 441 (App. Div. 2012) (EMT injured in car accident while returning from flu vaccine clinic suffered compensable injuries where she was encouraged to receive vaccination, it was provided at no cost, and she would not have gotten a flu vaccine otherwise); Maher v. Workers' Comp. Appeals Bd., 661 P.2d 1058 (Cal. 1983) (complications from tuberculosis treatment undergone by nurse's assistant after pre-employment test was positive for exposure to tuberculosis arose out of and in the course of employment); Suniland Toys & Juvenile Furniture, Inc. v. Karns, 148 So. 2d 523 (Fla. 1963) (injuries caused by typhoid inoculation offered to employees after they drank possibly contaminated tap water on employer's premises were compensable under the workers' compensation act).

37. *See, e.g.*, Martin v. Mars Mfg. Co., 58 N.C. App. 577, *cert. denied*, 306 N.C. 742 (1982) (injuries sustained at party sponsored and paid for by employer, which employer encouraged employees to attend and from which employer derived benefit, are compensable injuries); Chilton v. Bowman Gray Sch. of Med., 45 N.C. App. 13 (1980) (faculty member not entitled to workers' compensation benefits for injuries sustained while playing volleyball at annual radiology department picnic where department did not sponsor, did not require attendance at, and did not reap any definable benefit from picnic); Stiwinter v. Cmty. Newspapers, Inc., N.C.I.C. No. 165436 (1994) (Industrial Commission denied compensation for injuries suffered at staff picnic where employer did not sponsor or finance picnic, did not require attendance at picnic, and picnic was not held on employer's grounds).

Employers have sometimes argued that injuries resulting from vaccination should not be covered by workers' compensation because the vaccination is to protect the employee and thus for the employee's benefit rather than the employer's benefit. This argument is wrong for two reasons. First, in the context of human services and public health, each employee who chooses to undergo vaccination at the employer's encouragement does so in furtherance of that employer's public health, health care, or public safety mission and thus for the employer's benefit. And second, in North Carolina, the rule is that where an employee is injured while engaged in an activity for both the employee's own and the employer's mutual benefit, the injury is compensable as arising out of the employment.[38] The fact that the larger purpose of encouraging public employees to receive vaccinations is to protect the general public with which they come into contact does not affect compensability. The North Carolina courts have held that an injury that occurs while the employee is acting for the benefit of a third person can "arise out of" his or her employment, provided that the act appreciably benefits the employer.[39] Thus, complications from vaccinations required or offered by an employer "arise out of" employment within the meaning of the Act under any of the aforementioned standards and are thus compensable.

> *The commissioner has one further question. "Well, how about employees who contract bird flu on the job. Will they be entitled to workers' comp disability or death payments?" "Good question," the county attorney replies. "I don't think we're going to be able to pinpoint where employees who contract bird flu were infected. While it might have been on the job, especially if the employee works in public health, it might also have been in the supermarket or through infected family members."*

During an outbreak of any sort of contagious disease, it is very difficult, and often impossible, to determine where and how an individual became infected. Under the North Carolina Workers' Compensation Act, the burden is on employees to show that the condition from which they suffer arose out of and in the course of employment. Unless an employee can prove he

---

38. *See* Wall by Wall v. N. Hills Props., Inc., 125 N.C. App. 357, *discretionary review denied*, 346 N.C. 289 (1997).

39. *See* Roman v. Southland Transp. Co., 131 N.C. App. 571, 574 (1998), *aff'd*, 350 N.C. 549 (1999).

or she contracted the illness through employment—which seems unlikely where an infectious agent such as flu is widespread in the community—the employee's condition will not be compensable under the Act.

Monitoring Employee Health during a Public Health Emergency

When an emergency involves a communicable disease, such as a pandemic flu or a terrorist attack that spreads a lethal virus, public employers will want many employees—both those who have been exposed to the virus and those who are contagious—to stay home. They will also want to monitor the health of asymptomatic employees who report to work and to send home anyone who starts showing signs of illness during the workday. These are legitimate concerns, but employers must ensure that any monitoring measures they undertake comply with the ADA.[40] The ADA prohibits employers from requiring current employees to undergo a medical examination and from making inquiries of an employee as to whether the employee is disabled and about the nature or severity of the disability. Where a medical examination or inquiry is shown to be job-related and consistent with business necessity, the ADA provides an exemption from the prohibition.[41]

Even so, during a period when seasonal flu is active in the community, employers may ask employees if they are experiencing influenza-like symptoms, such as fever or chills and a cough or sore throat. Normal seasonal influenza, even at a pandemic level, is not a disability within the meaning of the ADA, and the EEOC has said that it will not consider questions like these

---

40. Title I of the Americans with Disabilities Act (ADA) and the EEOC regulations implementing Title I (29 C.F.R. Part 1630) apply to public employers with fifteen or more employees. See 42 U.S.C. § 12111(5)(A) (2002); 28 C.F.R. § 35.140(b)(1). The U.S. Department of Justice is charged with enforcing Title II of the ADA, and under its regulations, public employers with fewer than fifteen employees are subject to the employment-related requirements of Section 504 of the Rehabilitation Act of 1973, which prohibits discrimination against the disabled by recipients of federal funding, and the regulations promulgated thereunder. See 42 U.S.C. § 12132 (2002); 28 C.F.R. § 35.140(a). For the Rehabilitation Act of 1973, see 29 U.S.C. §§ 794 et seq. The original Rehabilitation Act regulations formed the basis of many of the regulations issued by the EEOC under Title I of the ADA.

41. See 42 U.S.C. § 12112(d)(4). See also 29 C.F.R. §§ 1630.13 and 1630.14, which address pre-employment medical inquiries and examinations.

to be disability-related and thus prohibited. The EEOC has further allowed that where a pandemic flu is one that causes severe symptoms, including death, questions about flu-like symptoms may be disability-related but are still allowable if the employer has a reasonable belief based on objective evidence that this severe form of pandemic flu poses a direct threat to other employees or the public.[42]

The EEOC's clear acknowledgment of the need to actively screen employees reporting to work during an outbreak of contagious disease is important, because the federal Centers for Disease Control and Prevention (CDC) advises employers to engage in active screening during pandemics: the CDC says that as the severity of an outbreak increases, employers should ask all employees at the beginning of the workday or the start of each shift whether they have fever or chills and cough or sore throat, the combination of symptoms that characterize flu, or symptoms characteristic of any other disease of which there is an outbreak.[43] An employer may not, however, ask an employee to disclose in advance if he or she has a compromised immune system or chronic health condition the CDC says could make him or her more susceptible to influenza complications because the response will likely disclose the existence of a disability. The ADA does not permit such an inquiry in the absence of objective evidence that pandemic symptoms will be a direct threat to the employee. Such evidence is completely absent before a pandemic occurs.

Taking Employee Temperatures

Taking an employee's temperature, harmless and straightforward as it may seem, constitutes a medical examination for ADA purposes.[44] Employers may not, therefore, take the temperature of an individual employee or screen any group of employees on a regular basis unless doing so is job-related and consistent with business necessity. One might argue that public health and public safety personnel who interact with the public on a regular and sometimes intimate basis are always at risk for infecting those with whom

---

42. *See Pandemic Preparedness in the Workplace.*

43. *See* Centers for Disease Control (CDC), *Guidance for Businesses and Employers to Plan and Respond to the 2009–2010 Influenza Season,* August 19, 2009 (hereinafter *Guidance for Businesses and Employers*), www.flu.gov/planning-preparedness/business/guidance.pdf.

44. *See Pandemic Preparedness in the Workplace,* at 6.

they come into contact, whether the infection is the common cold or a more serious contagious disease. Absent some extenuating circumstance, however, regular close contact with the public has never been recognized as the basis for finding a medical examination "job-related and consistent with business necessity."

The EEOC has recognized that SARS, bird flu, or new strains of seasonal influenza could reach pandemic proportions. It has therefore advised employers that in the event the CDC or state or local health authorities find that there are widespread severe flu symptoms within a community, employers may then measure their employees' body temperatures. Managers, directors, and human resources professionals must remember that they are not to make a determination about the seriousness of any outbreak of contagious disease on an ad hoc basis but must instead look to public health professionals to make that finding before embarking on a temperature-taking program.[45]

Employees Showing Symptoms of Illness

Employers may order employees showing signs of a contagious illness or who have been diagnosed with a contagious disease either to leave the workplace or to remain home. In the case of a potentially severe illness such as influenza, the CDC recommends that infected employees not report to work for at least seven days, even if their symptoms improve.[46] Ordering employees to stay at home when sick does not violate the ADA.[47]

There is no federal or state requirement that employers have a medical professional on-site to screen employees, take their temperatures, or decide an employee is showing symptoms of a communicable disease. Employers may designate whomever they wish to make the determination that an employee is too sick to remain in the workplace. They may designate someone in the human resources department to do the assessment, department heads, a paramedic or emergency medical technician from EMS, or simply someone whose workload is light at the time.

An employer may require employees to wear personal protective equipment during an outbreak of communicable disease. However, where an

---

45. *See Pandemic Preparedness in the Workplace*, at 6.
46. *See Guidance for Businesses and Employers.*
47. *See Pandemic Preparedness in the Workplace*, at 6.

employee with a disability needs an accommodation under the ADA (for example, non-latex gloves, or gowns designed for individuals who use wheelchairs), the employer must provide it unless doing so causes an undue hardship.[48] This is, however, unlikely, as most items of this kind are not very expensive.

Exclusion of Employees: What Is a Direct Threat?

An employer may not exclude a person with a disability from the workplace at any time—even in anticipation of a public emergency—unless that employee poses a direct threat to him- or herself or others. Under the ADA, a direct threat is "a significant risk of substantial harm to the health or safety of the individual or others that cannot be eliminated or reduced by reasonable accommodation."[49] A determination that an employee poses a direct threat must be based on objective, factual information, "not on subjective perceptions . . . [or] irrational fears" about a specific disability or disabilities.[50] Thus, the EEOC requires employers to consider four factors when determining whether an employee poses a direct threat: (1) the duration of the risk, (2) the nature and severity of the potential harm, (3) the likelihood that potential harm will occur, and (4) the imminence of the potential harm.[51]

Adopting a Communicable Disease Policy

To avoid confusion at a time already fraught with worry, employers should adopt a communicable disease policy. The policy should do the following:

- Define what constitutes a communicable disease or illness that can be transmitted through worker-related activities and is covered by the policy;
- Set forth the responsibilities of employees, their supervisors, and any in-house medical providers to advise the human resources department of the condition of sick or symptomatic employees, whether themselves or others;

---

48. *See Pandemic Preparedness in the Workplace*, at 7–8.
49. 42 U.S.C. § 12111(3).
50. *See Pandemic Preparedness in the Workplace*, at 7–8.
51. *See Pandemic Preparedness in the Workplace*, at 7–8.

- Explain what supervisors or co-workers are to do if, based on their own observations, they believe an employee is showing symptoms of a communicable disease covered by the policy;
- Set forth the employer's leave policies, even if they are set forth in other parts of the personnel policy;
- Specifically address how the employer will treat absences resulting from quarantines ordered by state or local health authorities;
- Set forth the employer's reasonable accommodation policies, which should incorporate both employee and employer obligations under the ADA; and
- Set forth the employer's return to work policy, including any requirement that an employee provide a fit-for-duty certification.

Two points are worth emphasizing here. First, employers will not want a communicable disease policy to cover every illness that spreads person-to-person. An ordinary cold, upper respiratory infection, or stomach virus can be handled through use of sick leave, as can typical cases of seasonal flu. Common contagious conditions like these will sometimes cause complications that require an employee to be absent for more than three days and to visit a health care provider for treatment two or more times. In such situations, employees will be eligible for leave under the federal Family and Medical Leave Act (FMLA). A communicable disease policy may be useful, however, where a contagious illness that is highly infectious and possibly lethal is widespread in the community. Examples of such illnesses include encephalitis, pandemic flu, bird flu or SARS, any of the hemorrhagic fever viruses such as Ebola or Marburg, smallpox, measles, tuberculosis, and antibiotic-resistant staph infection. The state Commission for Public Health has declared all of these illnesses to be "dangerous to the public health" and reportable to the local health director. The most recent list of reportable diseases as compiled by the commission may be found in the North Carolina Administrative Code at 10A N.C.A.C. 41A.0101. Employers will probably find it helpful to discuss with their local public health agency which diseases and conditions to include in a communicable disease policy.

Second, many employees with serious communicable diseases will be eligible for FMLA leave. In drafting a communicable disease policy, employers must make sure that their internal leave requirements are not inconsistent with FMLA requirements. The applicability of FMLA leave to public health emergencies is discussed below.

Handling Absences from Work during a Public Health Emergency

Government employees may be absent from work during a public health emergency for four primary reasons:

1. They are infected with a contagious illness and are unable to work.
2. They are caring for a family member who is infected with a communicable disease and needs care.
3. They are infected with or have been exposed to a contagious illness, but are functionally able to work.
4. They are well, but refuse to come to work because they are afraid that they will become ill if they leave home.

Employees absent from work because of the first or second reasons are likely to be covered by job-protected FMLA leave. Employees absent because of the third or fourth reasons will not be covered by the FMLA. Policies should be developed before an emergency develops that address absences for reasons three and four.

Absences Covered by the Family and Medical Leave Act

The federal FMLA requires employers to allow employees to take a set number of weeks off from work, without pay, in certain qualifying circumstances without losing their jobs or their health insurance coverage. The FMLA applies to all public employers.[52] This statement is somewhat misleading, however, because to be eligible for FMLA leave, a public employee must work at a worksite having at least 50 employees within a 75-mile radius.[53] Jurisdictions with fewer employees do not have to allow employees to take FMLA leave. FMLA-eligible employees get up to 12 workweeks of job-protected, unpaid leave during any 12-month period for any of the qualifying reasons set forth in the act and its regulations. FMLA leave would be available when (1) the employee has a *serious health condition*, as that term is defined in the act, that renders the employee unable to perform his or her job or (2) the

---

52. *See* 29 C.F.R. § 825.104(a).
53. *See* 29 C.F.R. § 825.110(a). In addition to the requirement that an employee work at a worksite with a minimum of 50 employees, to be eligible for Family and Medical Leave Act (FMLA) leave an employee must also (1) have a total of at least 12 months of service with the employer, although the 12 months need not be consecutive, and (2) have worked at least 1,250 hours during the last 12 months.

employee is needed to care for a spouse, child, or parent with a serious health condition.[54]

The FMLA regulations define a *serious health condition* as "an illness, injury or impairment, or physical or mental condition" that involves any period of incapacity that either

- requires an absence of more than three full, consecutive calendar days from work that also involves continuing treatment (that is, two or more times) by a health care provider within the first 30 days of incapacity or one visit that results in a regimen of continued treatment under the supervision of a health care provider or
- is connected with inpatient care.[55]

For FMLA purposes, a *period of incapacity* means an inability to work, attend school, or perform other regular daily activities due to the serious health condition, treatment for it, or recovery.[56] In a public health emergency, the disease or medical condition causing the emergency may also exacerbate chronic health conditions from which the employee already suffers, such as asthma, diabetes, or epilepsy, which are independent reasons for FMLA leave.[57]

Who Is Entitled to FMLA Leave?

A key element of the definition of a serious health condition is the requirement that an employee not be able to work or engage in daily activities. In a public health emergency involving a communicable disease, therefore, only employees who are so sick as to be unable to work or employees needed to care for sick family members qualify for FMLA leave. Employees who are functionally able to work, even if they are infected but symptom-free or have been exposed to the disease but remain symptom-free, do not qualify for FMLA leave. They may be required to work from home, if their job duties and the employer permit, or may use sick or vacation leave, if the employer's

---

54. *See* 29 U.S.C. § 2612(1); 29 C.F.R. § 825.112(a).

55. See 29 C.F.R. § 825.113(a).

56. *See* 29 C.F.R. § 825.113(b).

57. *See* 29 C.F.R. § 825.113(c). For a detailed discussion of FMLA leave, *see* DIANE M. JUFFRAS, EMPLOYEE BENEFITS LAW FOR NORTH CAROLINA PUBLIC EMPLOYERS (hereinafter EMPLOYEE BENEFITS LAW) 181–97 (UNC School of Government, 2009).

policies permit.[58] Employees quarantined pursuant to a state or local order but who are not ill also do not qualify for FMLA leave. Although a number of other states have enacted laws protecting employees from losing their jobs if they are unable to report to work because they are under an order of quarantine or isolation, North Carolina has not.[59]

What happens to FMLA leave when a natural disaster or public health emergency forces an organization to shut down for a period of time? Where a workplace is closed for less than an entire workweek, the employer may count the absences of an employee on FMLA leave against that employee's 12-week entitlement. Where the workplace is closed for a full week or more, however, the employer may not count the time during which the organization is shut down against an employee's FMLA entitlement. This is true even if it is obvious that the employee would be unable to come to work even if the organization were open. The FMLA regulations provide:

> If for some reason the employer's business activity has temporarily ceased and employees generally are not expected to report for work for one or more weeks (e.g., a school closing two weeks for the Christmas/New Year holiday or the summer vacation or an employer closing the plant for retooling or repairs), the days the employer's activities have ceased do not count against the employee's FMLA leave entitlement.[60]

Requesting Information about an Absent Employee's Medical Condition

When an employee is absent for more than three days, an employer may ask the employee to return an FMLA medical certification form signed by the employee's health-care provider. Although the FMLA does not require employers to obtain a medical certification or any kind of medical documentation at all, it does allow an employer to request certain information from the health-care provider to verify that an employee or family member has

---

58. For a detailed discussion of sick and vacation leave, see EMPLOYEE BENEFITS LAW, at 176–80.

59. See, e.g., DEL. CODE ANN. tit. 20 § 3136(6)(d); IOWA CODE § 139A.13A; KAN. STAT. ANN. § 65-129d; MD. CODE ANN., HEALTH–GEN. § 18-906; MINN. STAT. § 144.419b; N.J. REV. STAT. § 26:13-16; ME. REV. STAT. ANN. tit. 26, § 875: N.M. STAT. ANN. § 12-10A-16; UTAH CODE ANN. § 26-6b-3.3.

60. 29 C.F.R. § 825.200(h).

a qualifying condition.[61] While employers are entitled to obtain the information necessary to verify the need for FMLA leave, they do not have the right to know the actual diagnosis of the employee's condition.[62] The FMLA regulations permit employers to ask for relevant medical facts that support the employee's request for leave. In some cases the information they receive from the employee's treating physician may include a diagnosis.[63]

Requiring a Fitness-for-Duty Certification after a Public Health Emergency Absence

FMLA Leave. Employers may require employees returning from an FMLA leave for their own serious health condition to provide a medical certification that they are able to return to work and resume their duties. Such a certification, however, must be requested pursuant to a policy or practice that is uniformly applied to all similarly situated employees. In other words, an employer may not ask for FMLA medical certification ad hoc or for just one employee about whom management has suspicions. "Similarly situated" might mean all employees, all employees with the same occupation or job classification, all employees within specified departments, or all employees who have been on leave for two weeks or more. "Similarly situated" could also mean all employees on FMLA leave during a public health emergency.

An employer may not require a return-to-work certification unless it has given employees written notice of the requirement at the time the employer responds to the initial request for leave.[64] An FMLA return-to-work certification is less detailed than that which may be required to support an employee's request for leave, although the regulations do allow an employer to require that a health-care provider certify that the employee can perform a list of duties specific to the employee's job.[65] To require that the return-to-work certification address the employee's ability to perform the essential functions of his or her job, an employer must give the employee a list of essential job functions at the outset, when responding to the initial request

---

61. *See* 29 C.F.R. § 825.305.

62. See the Summary of Comments prefacing publication of the Family and Medical Leave Act Final Rule, 29 C.F.R. Part 825, at 73 Fed. Reg. 68014 (Nov. 17, 2008) (comment on 29 C.F.R. § 825.306).

63. *See* 29 C.F.R. § 825.306 and Appendix B to 29 C.F.R. Part 825.

64. *See* 29 C.F.R. §§ 825.300(d), 825.312(d).

65. *See* 29 C.F.R. § 825.312(b).

for leave. An employer may not add this requirement as an afterthought at the time of return to work. Although public employers are likely to be overwhelmed during a public health emergency, this may be one of the situations in which obtaining a return-to-work certification is most important. Employers must therefore handle requests for FMLA leave carefully so that if they wish to require return-to-work certifications or certifications that address the employee's ability to perform essential job duties, they make a timely request. An employee who does not provide a required return-to-work certification is not entitled to reinstatement.[66] But employers should note that as a practical matter, doctors and other health-care professionals may be too busy during and immediately after a communicable disease outbreak to provide fitness-for-duty documentation.[67]

Fitness-for-Duty Certifications for Health-Related Absences That Do Not Qualify for FMLA Leave. Some employees may not report to work for three days but may return on the fourth, keeping their absence from qualifying as a serious health condition under the FMLA. Others may have already exhausted their FMLA entitlement for that year. Still others may not have worked for the employer for the requisite number of months and hours to be eligible for FMLA leave. Employees of public entities with fewer than 50 employees are not eligible for FMLA leave under any circumstances. May a public employer ask employees in these categories for a fitness-for-duty certification after an absence due to a communicable disease? The answer is yes, notwithstanding the ADA's prohibition against asking current employees questions about possible disabilities. According to the EEOC, fitness-for-duty certifications are permitted under the ADA either because normal seasonal viruses are not disabilities and the certification would not, therefore, be disability-related or, if the outbreak of disease were truly severe, employers would be justified in making what might be a disability-related inquiry because protecting other employees would be job-related and consistent with business necessity.[68]

---

66. *See* 29 C.F.R. §§ 312(e); 313(d)(3).
67. *See Pandemic Preparedness in the Workplace*, at 9.
68. *See Pandemic Preparedness in the Workplace*, at 9.

Short-Term Disability and Workers' Compensation
Benefits during FMLA Leave

Both short-term disability plans and workers' compensation insurance exist to provide ill or injured employees with replacement income, but they apply in different circumstances. The purpose of short-term disability is to provide income replacement when an employee cannot work due to an illness or injury that is not work-related. The purpose of workers' compensation insurance is to provide income replacement and cover medical expenses incurred when an employee becomes ill or injured in the course of employment. Employees who become ill or injured outside of work during a natural disaster or public health emergency may, therefore, qualify for short-term disability benefits if their absence from work is more than a few days. Short-term disability plans are a discretionary benefit and the terms of the plan will vary from jurisdiction to jurisdiction. In addition, because the benefit is not required by law, local government employers may put whatever conditions they choose on its use. Some employers, for example, do not allow employees to be eligible for short-term disability until all accrued sick leave has been used. Others do not allow employees to become eligible for short-term disability until they have completed a probationary period. In assessing whether employees who are unable to come to work because of a medical condition acquired outside of work are eligible for short-term disability benefits, each employer must consider carefully the terms of the plan itself and the provisions set forth in its individual personnel policies.

Workers' compensation insurance, on the other hand, is designed to provide income replacement for persons injured on the job. An employee who claims to have caught pandemic flu or another communicable disease on the job would probably not be eligible for workers' compensation benefits. As discussed earlier in the section on required vaccinations, one of the elements a workers' compensation claimant must show is that the injury was sustained in the course of the employment—that is, the injury occurred while the employee was working.[69] In the case of a disease prevalent in the community, it would be virtually impossible for employees to prove they contracted the condition at work. The more severe the symptoms of the disease and the higher the death toll, the more fearful people become of exposure.

---

69. *See, e.g.,* Cole v. Guilford Cnty., 259 N.C. 724 (1963); Taylor v. Twin City Club, 260 N.C. 435, 438–39 (1963); Fulcher by Wall v. Willard's Cab Co., 132 N.C. App. 74, 81–82 (1999); Sisk v. Tar Heel Capital Corp., 166 N.C. App. 631, 635 (2004).

Could a first responder, whose contact with infected persons would likely be significant, file a successful workers' compensation claim based on emotional distress due to the work or a fear of exposure to the disease through work? Possibly, but the author has not seen any North Carolina cases recognizing a basis for such a claim.

Employees Who Are Absent but Able to Work

In addition to employees who are absent and unable to work during a public health emergency because of their own or a family member's serious health condition, some employees will stay home because they are either ill (but not incapacitated) or they have been exposed to others who are ill. Employees with disabilities that put them at high risk for complications of pandemic influenza or another communicable disease in circulation may also need to remain at home to reduce their chances of infection. Yet another group of employees who are absent from work will be those afraid to leave their homes or come to work because of a fear of contagion. All of these employees are functionally able to work. Employees who have confirmed cases of disease will probably qualify for FMLA leave if they are ordered to stay home by their health care provider for more than three days (reasonably likely in the case of a pandemic) and they seek treatment from their physician two or more times, or, as may be more common, undergo a regimen of treatment supervised by their provider. This regimen of treatment may be no more complicated than taking antibiotics, antivirals, or medication designed to lessen the symptoms of the disease. Employees at risk for complications due to a disability, on the other hand, will not normally qualify for FMLA leave. Persons who do not come into work because they are fearful of exposure cannot qualify for FMLA leave, as they do not have a serious health condition.

Employers will need to decide whether to allow these three types of employees to work from home and determine the positions for which such an arrangement will be possible. Employers are not required to allow employees who are able to work to work from home. Employees who do work from home, however, cannot have that time counted against their FMLA entitlement. Because FMLA leave is unpaid (unless run concurrently with sick, vacation, or comp time), some employees may feel they must return to work as soon as possible because they cannot afford to go without a paycheck. Allowing exposed or asymptomatic employees, as well as those still recovering, who are functionally able to work to telecommute will lessen the risk to other employees reporting to the regular workplace (as will allowing generous use of sick and vacation leave, advancing sick leave, creating leave banks, and assisting with applications for short-term disability).[70] Employees at high risk for developing complications from a disease that is widespread in the community should be allowed to work from home as a form of reasonable accommodation.[71]

Informing Other Employees of a Co-worker's Illness

In the case of a widespread outbreak of a contagious disease, an employer may feel an obligation to other employees to inform them of a co-worker's illness. The law, however, does not permit an employer to do so. Both the ADA and the FMLA require employers to keep medical and genetic information about an employee on separate forms in separate medical files and to treat such information as a "confidential medical record."[72] Although an employer may not disclose individually identifiable information about an employee's medical condition to co-workers, it may nevertheless inform them, without identifying the particular employee, that they may have been exposed to the illness in question and advise them to monitor themselves for symptoms of the illness and stay home if they develop symptoms.[73]

Medical information may be given to supervisors only if they need to have it. Medical information may be relevant to department heads and supervisors if the employee will need a reasonable accommodation as a result of an illness and these members of the management team will be involved in devising or implementing an accommodation.[74]

---

70. In the case of a severe strain of influenza virus, for example, the CDC recommends that ill persons not come to work or travel and remain at home for at least seven days, even if symptoms resolve sooner and whether or not they use antiviral medications. See *Guidance for Businesses and Employers*.

71. *See Pandemic Preparedness in the Workplace*, at 7.

72. *See* 42 U.S.C. § 12112(d)(3)(B); 29 C.F.R. § 1630.14(b)(1). *See also* 29 C.F.R. § 1635.9.

73. See www.flu.gov/faq/workplace_questions/equal_employment/index.html#PrivacyIssues, at 8.

74. *See* EEOC, *ADA Enforcement Guidance: Preemployment Disability-Related Questions and Medical Examinations*, Notice Number 915.002 (Oct. 10, 1995), at 20–21, www.eeoc.gov/policy/docs/medfin5.pdf.

Employees Who Refuse to Work

In addition to employees who refuse to leave home during a public health emergency, some first responders and public health employees may refuse to handle infected patients, especially where the mortality rate from the infection is high. Under the Occupational Safety and Health Act of North Carolina,[75] which incorporates by reference the federal Occupational Safety and Health Act and its regulations,[76] employees may refuse to work in the face of a real danger of death or injury:

(b)(1)   On the other hand, review of the Act and examination of the legislative history discloses that, as a general matter, there is no right afforded by the Act which would entitle employees to walk off the job because of potential unsafe conditions at the workplace. Hazardous conditions which may be violative of the Act will ordinarily be corrected by the employer, once brought to his attention. If corrections are not accomplished, or if there is dispute about the existence of a hazard, the employee will normally have opportunity to request inspection of the workplace pursuant to section 8(f) of the Act, or to seek the assistance of other public agencies which have responsibility in the field of safety and health. Under such circumstances, therefore, an employer would not ordinarily be in violation of section 11(c) by taking action to discipline an employee for refusing to perform normal job activities because of alleged safety or health hazards.

(2)   However, occasions might arise when an employee is confronted with a choice between not performing assigned tasks or subjecting himself to serious injury or death arising from a hazardous condition at the workplace. If the employee, with no reasonable alternative, refuses in good

---

75. The federal Occupational Safety and Health Act (OSHA) excludes state and local government employees from its coverage. They are, however, covered by the Occupational Safety and Health Act of North Carolina, which incorporates the federal OSHA rules and standards except in the few places where the N.C. Department of Labor has issued its own rules. See 29 U.S.C. §§ 652(5) and (6). See G.S. Chapter 96, Article 16, for the Occupational Safety and Health Act of North Carolina.

76. See G.S. 95-131(a); 29 C.F.R. Part 1910.

faith to expose himself to the dangerous condition, he would be protected against subsequent discrimination. The condition causing the employee's apprehension of death or injury must be of such a nature that a reasonable person, under the circumstances then confronting the employee, would conclude that there is a real danger of death or serious injury and that there is insufficient time, due to the urgency of the situation, to eliminate the danger through resort to regular statutory enforcement channels. In addition, in such circumstances, the employee, where possible, must also have sought from his employer, and been unable to obtain, a correction of the dangerous condition.[77]

The standard set out in the regulation quoted above will be difficult for employees working in public safety or public health to meet in a public health emergency, given that their work regularly involves a greater risk of exposure to contagions and contaminants than that faced by other government employees or the public. The author has not found any reported cases at the state or federal court level in which interpretation of this regulation has been at issue.[78] In most cases, then, an employer may discipline or discharge an employee who refuses to work in an emergency.

---

77. See 29 C.F.R. § 1977.12(b).

78. For some health care professions, professional standards may dictate the circumstances under which providers may decline to provide services. Interestingly, the Model Public Health Emergency Powers Act, which has been adopted by a number of states but not North Carolina, contains a provision granting public authorities the power "to require in-state health care providers to assist in the performance of vaccination, treatment, examination, or testing of any individual as a condition of licensure, authorization, or the ability to continue to function as a health care provider in this State." See Center for Law and the Public's Health, Georgetown and Johns Hopkins Universities, *Model State Emergency Health Powers Act Discussion Draft*, Section 608 (2001), www.publichealthlaw.net/MSEHPA/MSEHPA2.pdf. *See also* Carl H. Coleman, *Beyond the Call of Duty: Compelling Health Care Professionals to Work During an Influenza Pandemic*, 94 IOWA LAW REVIEW 1–46 (2008), www.law.uiowa.edu/documents/ilr/coleman.pdf.

Employees Covered by the State Personnel Act

Local health care, social services, mental health care, or emergency management employees who are covered by the State Personnel Act (SPA) and who refuse to report to work during an emergency may be disciplined on grounds that they have engaged in insubordination (a form of personal misconduct) by refusing to follow a reasonable and direct order from a supervisor. There are no provisions for suspending the SPA in an emergency. This means, among other things, that employees who engage in misconduct, including insubordination consisting of refusals to report to work or to perform certain assigned duties, must be granted their due process rights of notice and opportunity to be heard before any disciplinary action may be taken against them.

Conclusion

Despite their best efforts, local government employers will never be able to prepare for every contingency that might arise in an emergency. The sheer number of types of possible emergencies and disasters is too large, as is the potential variation in the scope and seriousness of each. Nevertheless, a local government can take steps to address emergency situations before they occur.

First, local government employers can plan how they and their employees should respond to certain circumstances likely to arise whatever the type of emergency. These employers should:

- develop an employee evacuation plan that takes into account disabilities and other limitations individual employees may have;
- become compliant with the National Incident Management System and develop backup chains of command to be implemented in case of the absence or incapacity of key members of the management team, alternate command chains for different types of emergency responses, and lists of essential employees for different kinds of emergencies;
- educate their employees about the backup chain of command and inform those whose services will be essential that they will be needed and expected to report to work in a given situation;
- consider how to handle any increased rate of absenteeism in departments that will be crucial to an emergency response in different circumstances; and

- offer those employees who will perform natural disaster cleanup work a pre-exposure medical screening so that the employer may better track any reactions to material encountered in the course of the work.

Local government employers should also make a number of technical and policy decisions in advance of an emergency situation, including whether its information technology infrastructure can support greater than usual numbers of employees accessing the IT system remotely and whether it wants or will allow essential and nonessential employees to telecommute. If a local government anticipates that some employees will work remotely, it should develop a policy addressing whether and how the work schedule will change, how supervisors will know an employee is working at any given time, and how employees will record remote work time.

Because all emergency responses, whatever their nature and size, put a strain on local government finances, the county or city manager, the human resources and finance departments, and supervisors at every level should be familiar with how to apply the Fair Labor Standards Act (FLSA) in a disaster. A local government can apply for reimbursement from the Federal Emergency Management Agency for overtime employees work as a result of the disaster. If the local government anticipates that its exempt employees will also work additional hours and it wishes to compensate them for that time, it should adopt a policy establishing the emergency circumstances and rate of compensation for such exempt employee work in advance. This is not a guarantee of reimbursement, but without such a written policy in place in advance of the disaster, the federal government will not even consider an application for reimbursement.

The FLSA mandates that employees must be paid on the next regular payday—even in a disaster situation. Employers can and should acquire network redundancy and backup access to the software needed to run payroll and transfer wages and taxes to the applicable accounts in an emergency. They must also have available off-site a backup of all the supporting information needed to run payroll, including Social Security numbers for tax reporting purposes, bank account numbers, and bank routing information.

Several issues are unique to public health emergencies. Because many of them involve human-to-human spread of serious illnesses, employers should take the initiative and adopt a communicable disease policy before an outbreak of some sort occurs. The policy should address, among other things, the responsibility of a sick employee to notify the employer of his or her condition, the responsibility of supervisors or co-workers to report an employee whom they believe is showing symptoms of a serious communicable disease covered by the policy, how the employer will treat absences resulting from quarantines ordered by state or local health authorities, and the employer's return-to-work policy.

Because public health emergencies will frequently involve workplace screenings for symptoms of illness and requests for family and medical leave, employers should instruct everyone on the management team and all supervisors in the requirements of medical confidentiality under the Family and Medical Leave Act, the Americans with Disabilities Act, and those provisions of the North Carolina General Statutes addressing public employee personnel privacy.

Taking the steps set forth above will not lessen the severity of emergencies that are the result of natural disasters, criminal activities, or an outbreak of serious communicable disease, but it will make responding to these situations somewhat easier and less chaotic for city and county governments.